W. T. Sharp.

MODERN GEOMETRY

MODERN GEOMETRY

AN INTEGRATED FIRST COURSE
SECOND EDITION

CLAIRE FISHER ADLER, PH.D.

PROFESSOR OF MATHEMATICS
C. W. POST COLLEGE OF LONG ISLAND UNIVERSITY

McGraw-Hill Book Company

NEW YORK ST. LOUIS SAN FRANCISCO TORONTO
LONDON SYDNEY

MODERN GEOMETRY

AN INTEGRATED FIRST COURSE

Library of Congress Catalog Card Number 66-25471

00421

1234567890 MP 7432106987

preface

This is the second edition of a text whose favorable reception has encouraged the author to undertake the considerable task of revising and updating the original. Experience, the thoughtful comments of reviewers and teachers, and the passage of time have shown what could be condensed and what was needed to round out and modernize the original text. Mathematics is a living, growing subject, and as knowledge advances, mathematical developments tend to outflank the carefully prepared positions taken by authors.

The revision is directed to the needs of present and future students, and its main objectives are:

1. To acquaint the reader with the great discoveries and developments that followed Euclid
2. To meet the specific needs of prospective teachers of the secondary curriculum
3. To round out the geometric background of specialists in other branches of mathematics and other disciplines

The prerequisites are a reasonably good knowledge of the geometry of the secondary curriculum and some acquaintance with college algebra, trigonometry, and elementary analytic geometry.

Although the book has been entirely rewritten, the most noteworthy changes appear in Parts I and III.

Chapter 1 reflects the growing demand for logical rigor in today's mathematics. Although parts of this material overlap presentations in freshman college courses, the author feels that the geometric viewpoint, modern terminology, and symbolism need reviewing.

Much of Chap. 1 is a preparation for Chap. 2 in which axiomatic systems are illustrated and analyzed and flaws in Euclid are pointed out. The chapter also brings out the fact that two geometries will have different sets of axioms, just as two games will have different sets of rules.

The introduction to non-Euclidean geometry in Chap. 3 is informal.

Less attention is given to non-Euclidean geometry than was done fifty years ago, but teachers need at least a speaking acquaintance with this system.

The main change in Part II, a formal approach to projective geometry, is the insertion of proofs that were omitted in the original text. This classical synthetic approach to projective geometry, and the magnificent theory that has resulted, is compared later with the algebraic treatment of the same theory. The aesthetic appeal of pure geometry cannot be denied.

Part III is now a presentation in the modern style, using vectors and matrices as basic tools. This is both a *geometric approach to linear algebra and an algebraic approach to projective geometry.* A central theme is that of linear transformations and their use in classification and unification theories.

An introduction to the Erlanger program for classifying geometries and a final chapter on complex metric projective geometry bring the separate developments into one organic whole.

Early foundations of Euclidean geometry in Appendix A and Hilbert's axioms in Appendix B, as well as an extensive up-to-date bibliography, are inserted for the convenience of the reader.

Other special features of the text are closing remarks and a selected reading list at the end of each chapter. The closing remarks summarize highlights of the chapter, and the reading list supplies valuable source material for the specialists and/or all who wish to supplement the text.

There is more material here than can be covered in a one-semester course that meets three hours a week. But it is possible, by careful selection of the material, to use the text for either a *basic elementary course* or an *advanced course.*

Suggested selections for these courses are Chaps. 1 to 5 and selections from Chaps. 6 to 8 for persons with a modest mathematical background; Chaps. 5 to 11 for upperclass mathematics majors.

In addition to the acknowledgments made in the original edition, I wish to express my thanks to Dr. Sylvan Wallach for his suggestion that linear algebra concepts be introduced into the course and also to Mrs. Lorette Riddick for her gracious secretarial assistance in preparing this edition for publication.

Claire Fisher Adler

contents

PART ONE

FOUNDATIONS AND INTRODUCTION TO NON-EUCLIDEAN GEOMETRY

1 symbolic logic and set theory preliminaries

Because of the vast amount of material now classified as geometry and because of the hundreds of years in which this material has been accumulating, it is impossible in discussing modern theories to do justice to all phases of the science. Yet, to specialize is to lose the perspective necessary for gaining a proper understanding of the nature, the scope, and the importance of geometry in the twentieth century. For this reason, it has been deemed advisable to consider first the underlying structure or foundation of geometry. Specialization will then come later.

Attention is focused first on logic, the science used by Euclid and his successors in the development of his famous system. Euclidean geometry is but one of a hierarchy of geometries, each based on its own set of axioms and each having its own special application to the physical universe. Projective geometry and the classical non-Euclidean geometries are other famous systems to be investigated here.

Aristotle (385–322 B.C.), the great Greek philosopher, was one of the first to investigate the type of reasoning in which to employ his own description: "Certain things being stated, something other than what is stated follows of necessity from their being so." His investigations represent the beginnings of classical logic.

Like geometry, the ancient science has been modernized. Classical logic, so brilliantly and systematically laid, now forms only a small part of a much more elaborate system called symbolic logic.

Since the publication in 1910 of the monumental three-volume work "Principia Mathematica" by Bertrand Russell and A. N. Whitehead, symbolic logic has been a vigorously growing science engaging the talent and skill of some of the world's most gifted young scientists.

In the following introduction to the new science, the reader will find the minimum amount of modern equipment he is expected to bring with him to the absorbing study of geometric systems. Logic as a science in itself has been subordinated to logic as a tool—the tools in this case being the new terminology, the new symbolism, and the new symbolic methods that have brought precision and clarity into an ancient and respected science.

A preliminary discussion of reasoning in the inductive sciences will motivate and add interest to many of the abstract discussions that follow.

1.1 Reasoning in the inductive sciences

What man has accomplished through his ability to reason has been told over and over again, and yet it bears repetition. Reasoning of some kind or other is a perfectly simple, natural process carried on by normal people in the course of their daily activities. There is, for example, a *reasoning by analogy,* such as is done by an individual who notes the number of air fatalities before deciding whether or not to travel by plane. Obviously he is reasoning about his chances of survival on the basis of the number of airplane accidents which have happened in the past. Again, a person hearing a weather report of rain for the day will prepare for this eventuality on the assumption that the reports have been reliable in the past and will continue to be so.

The process may be more formalized. A doctor, for instance, arrives at a diagnosis of a specific disease by a careful, systematic investigation of all factors attendant upon the disease. He notes all symptoms, however trivial, excluding none until they have been proved irrelevant. He classifies, examines, and combines pertinent facts until he finally reaches the diagnosis which enables him to effect a cure. Reasoning of this kind is *inductive.*

Often, to verify a conclusion reached by inductive reasoning, the investigator makes repeated experiments. Sometimes they are carried on for a period of years, as was the case in establishing laws of astronomy, laws of heredity, and the famous law of falling bodies: $s = \frac{1}{2}gt^2$, where s is the distance in feet a body falls in t seconds and g is the gravitational constant whose value is approximately 32.2 feet per second.

Many other examples may be cited. There is, for instance, the constantly recurring phenomenon of the sun's rising and setting each day, from which it has been concluded that the sun will rise and set every day in the future. This does not mean, however, absolute certainty that the sun will rise tomorrow, or a week, or a year, or a million years from now. Conclusions reached by induction are only statements of what is more or less likely to happen. If an exception is ever found, either the conclusion is discarded, or laws of probability are used to determine its reliability for future prediction.

Inductive reasoning reaches its highest goal in the inductive sciences: physics, chemistry, biology; and the behavioral sciences: economics, political science, psychology, and sociology, to mention a few. The importance of these sciences in providing quantitative descriptions of natural phenomena and laws of the physical universe cannot be underestimated. The array is impressive. Satellites now encircling the earth are the result of Newton's earlier suggestion that projectiles shot out horizontally and with large velocities from the top of a high mountain would encircle the earth. High-speed computers have emerged from early calculating machines. Deep and intensive research in disease has lowered the death rate in infancy and has extended the life span. Today space craft is racing toward the moon at a speed in excess of 2,000 miles per hour with cameras transmitting photographs of the moon's surface back to earth. Man's rendezvous with the moon is predicted by 1970, and hopes are high for unlocking mysteries of the ages.

But not all the reasoning being used in these experiments is inductive. Aiding and supplementing the inductive processes is the all-powerful deductive reasoning upon which ultimately rests the whole of mathematics. How modern symbolic methods are used to simplify, clarify, and extend the process is explained in what follows.

1.2 Simple statements and classical laws of reasoning

Some attention is given first to the syntax of mathematical language and to the conventions regarding the truth or falsity of certain combinations of words called *statements*.

By a statement or proposition is meant here a meaningful sentence to which the notion of truth or falsity may be attached. "Today is Monday" is a statement which is factually true or false depending on a number of factors, such as where and on which day of the week the statement is made. But, statements like "Oh that I were a bird," or "Shut the door," or "X is an integer" are not statements in the sense just indicated, for one cannot say of them that they are either true or false.

Assuming the customary meaning of truth and falsity in a factual sense, few will deny that if the statement:

Today is Monday.

is true, its denial:

Today is not Monday.

is false. There seems to be no other possibility, and yet L. E. J. Brouwer, a famous Dutch mathematician of the twentieth century and a leader of the intuitionist school of thought, has raised some question about this matter when applied to infinite sets ([131] Chap. 10).[1]

Absurd as it may seem, not all simple declarative sentences can be either affirmed or denied. This is true, of course, of statements such as "World War I was fought to preserve democracy," whose truth or falsity might possibly be

[1] Numbers in brackets refer to entries in the Bibliography.

a matter of opinion or belief. It is true also of other types of statements. Suppose, for example, a man says:

I am lying.

Is his statement true? If so, he is lying, and his statement is false. Is his statement false? If so, he is lying, and his statement is true.

Again, suppose that Mr. Smith is a small-town barber who shaves those men and only those men of the town who do not shave themselves. Which of the following statements is true?

Mr. Smith shaves himself.

Mr. Smith does not shave himself.

Note the dilemma you are in if you attempt to answer this question. If you affirm the first statement and deny the second, you are admitting that Mr. Smith is shaving somebody who shaves himself, contrary to the hypothesis; if you reverse your opinion, affirming the second and denying the first, Mr. Smith is not shaving somebody who does not shave himself, and in this case, too, the hypothesis is contradicted.

Again, can you say which of the following statements is true?

$2^{\sqrt{2}}$ is a rational number.

$2^{\sqrt{2}}$ is not a rational number.

It will be recalled that a number N is rational if there exist two integers m, n such that $N = m/n$. As late as the year 1941, no one had found two integers m, n satisfying the condition $2^{\sqrt{2}} = m/n$ but neither had anyone proved that such integers did not exist. It was therefore impossible to affirm, deny, prove, or disprove either statement, up to the year just mentioned. The question has since been settled ([95] no. 9, p. 45).

Fortunately, classical logic avoids controversial questions of such a nature by assuming in advance of an argument that any given statement is either true or false. This is one of the three cornerstone laws of classical Aristotelian logic, called, respectively, the law of *identity,* the law of the *excluded middle,* and the law of *noncontradiction:*

1 *A thing is itself.* (Identity)
2 *A statement is either true or false.* (Excluded Middle)
3 *No statement is both true and false.* (Noncontradiction)

Long-accepted patterns of reasoning are woven around these three laws, but, strange to say, they are definitely not usable in vast regions of modern mathematics. In 1912, Brouwer challenged the second law, and a few years later Count Alfred Korzybski, a Polish-American logician, challenged the first. The third law, which deals with the consistency of a system, is discussed in detail in the next chapter.

1.3 Compound statements: conjunctions, disjunctions, conditionals, equivalences, and their symbolic representations

Statements may be simple or compound. A statement is *simple* if it can be analyzed into a single predicate and a noun about which the predicate says something, such as Aristotle's first law (Sec. 1.2), or the more homely illustration:

The house is white.

Let simple statements, such as these, be denoted by the letters p, q, and their denials by $\sim p$, $\sim q$, where $\sim$ is the denial symbol.

By the denial of statement p is meant the statement which is false (F) when p is true (T), and true when p is false.

A statement is *compound* if it is made up of two or more simple statements joined by a connective of some kind or other.

Four connectives used extensively by mathematicians are:

and *or* *if-then* *if and only if*

and respective symbols for the connectives are:

$\wedge$ $\vee$ $\rightarrow$ $\leftrightarrow$

Thus, if p denotes the simple sentence:

p: A statement is true.

the symbolic representation of Aristotle's second law is

$$p \vee (\sim p)$$

Table 1 lists four basic compound statements, their symbolic representations, and special names given to them in the literature.

p and q	$p \wedge q$	Conjunction
p or q	$p \vee q$	Disjunction
if p then q	$p \rightarrow q$	Conditional
p if and only if q	$p \leftrightarrow q$	Equivalence

TABLE 1

They appear in elementary geometry theorems, such as:

A triangle is isosceles and *equilateral.*	(Conjunction)
An integer is even or *odd.*	(Disjunction)
If two straight lines intersect, then *the vertical angles are equal.*	(Conditional)
A triangle is equilateral if and only if *it is equiangular.*	(Equivalence)

Brackets, braces, and other signs of grouping play much the same role in symbolic logic that punctuation marks play in the formation of sentences. For example, the expression

$$\sim(p \wedge \sim p)$$

which denies the conjunction $p \wedge \sim p$ is the symbolic representation of Aristotle's third law: "No statement is both true and false."

A change of position of parentheses in the expression

$$(p \vee q) \rightarrow \sim r$$

to

$$p \vee (q \rightarrow \sim r)$$

gives an entirely different statement.

Thus, if p, q, r are the statements

$$p\colon a = b \qquad q\colon a < b \qquad r\colon \angle A > \angle B$$

the first symbolic expression reads: "If $a \leq b$, then $\angle A$ is not greater than $\angle B$," and the second reads: "Either $a = b$, or if $a < b$, then $\angle A$ is not greater than $\angle B$."

Exercise 1.1

1 Write each of the following statements in symbolic form, and state which is (a) a conjunction, (b) a disjunction, (c) a conditional.
   a. A complex number is either real or nonreal.
   b. Line L is parallel to line M and perpendicular to line N.
   c. A triangle may be isosceles or equilateral.
   d. If p implies q, then p is false or q is false.
   e. An integer x is either even or odd.
   f. If triangle ABC is equilateral, it is isosceles and equiangular.

2 Translate each of the following into words, after substituting simple geometric statements for p, q, and r:

a. $p \rightarrow (q \vee r)$
b. $(p \rightarrow q) \vee r$
c. $(\sim p) \vee (q \wedge r)$
d. $\{p \wedge (\sim q) \vee q \wedge (\sim p)\} \rightarrow (p \vee q)$

1.4 Truth tables

The agreement that any statement, simple or compound, must be either true (T) or false (F), but not both, poses some agreement about *truth values for compound statements.*

If a compound statement S involves only the two simple statements p and q, then S has a truth value for each of the four possible pairs of truth values of p and q: (T, T), (T, F), (F, T), (F, F), and a simple device for tabulating them is a truth table such as Table 2.

		(1) *conjunction*	(2) *disjunction*	(3) *conditional*	(4) *equivalence*
p	q	$p \wedge q$	$p \vee q$	$p \rightarrow q$	$p \leftrightarrow q (p \equiv q)$
T	T	T	T	T	T
T	F	F	T	F	F
F	T	F	T	T	F
F	F	F	F	T	T

TABLE 2

Current usage has dictated many of the entries in the table. Column 1, for example, shows that the *conjunction* $p \wedge q$ is false in all cases except the one in which p and q are both true. This is the usual meaning for the connective "and."

Column 2 for the *disjunction* $p \vee q$ needs some additional explanation. Usually the word "or" is used in the exclusive sense meaning thereby that the occurrence (truth) of one thing, p, prevents the occurrence (truth) of another, q, as in the statement:

I shall go or I shall stay.

I can't do both. But often the precise meaning of the word is in doubt. For example, the statement:

I shall buy a black or red tie.

may mean the purchase of both. If so, the word "or" is being used in the nonexclusive "and/or" sense of legal documents, in which the statement "p

or q" includes the possibility of both p and q being true. In this text the word "or" is being used in the nonexclusive sense as shown in column 2 of Table 2.

Entries for the *conditional* $p \rightarrow q$ in column 3 of the table must be carefully noted. What does it mean, for example, to say:

If it is raining, then John is studying.

Here current usage is only a partial guide. Most persons will agree that the statement is false if there is a single rainy day on which John is not studying. On the other hand, nothing is to be read into the statement about John's study habits on a sunny day. On such a day John might or might not be studying. In other words, when p is false, q may be either true or false. The completely arbitrary decision was made, therefore, to label the conditional $p \rightarrow q$ true when p is false regardless of the truth value of q. Thus the conditional $p \rightarrow q$ is true for every combination of truth values of p and q except the one in which p is true and q is false.

Particular note is made now of the extension of the familiar cause and effect relationship of "if p, then q" statements, where p is customarily called the *antecedent*, and q the *consequent*. In the new interpretation of the conditional, p and q need not be related at all, as in the statement:

If the sun is shining today, $1 + 1 = 3$.

To attach a truth value to such a statement seems weird, but nevertheless it is done. The statement is false if the sun is (actually) shining and is true if the sun is not shining. This is so because the consequent $1 + 1 = 3$ is false; the conditional $p \rightarrow q$ is then true only when p is false, or in other words, if it is true that the sun is *not* shining.

An explanation of the words "converse" and "contrapositive" is in order at this point.

When p and q are simple statements, the conditional $p \rightarrow q$ has for its *converse* the conditional $q \rightarrow p$ and for its *contrapositive* the conditional $\sim q \rightarrow \sim p$, from which it is seen that an interchange of antecedent and consequent in the conditional $p \rightarrow q$ gives its converse $q \rightarrow p$, and the latter with denial signs inserted before q and p gives the contrapositive $\sim q \rightarrow \sim p$ of $p \rightarrow q$.

For example, the elementary theorem: "If a quadrilateral is a rhombus, its diagonals are perpendicular" has for its converse the false statement: "If the diagonals of a quadrilateral are perpendicular, the quadrilateral is a rhombus."[1]

[1]See Sec. 1.7, Example 2.

Again, the theorem: "If in triangle ABC, $\angle A = \angle B$, then $a = b$" has for its contrapositive the theorem: "If in triangle ABC, $a \neq b$, then $\angle A \neq \angle B$."

There is not universal agreement as to what constitutes a converse of a statement $A \to B$ if A and B are compound statements ([43] pp. 71–76).

Like the conditional, the *equivalence* $p \leftrightarrow q$, or preferably $p \equiv q$, deserves special note, so universally used and misused is the term.

Equivalence is often used loosely to denote equality, as when one speaks of two persons having equal power; and often to classify, as when one speaks of two persons having equal rank. Again, in algebra, two equations are said to be equivalent if their solution sets are identical; and in Euclidean geometry, the equiangular and equilateral properties of a triangle are said to be equivalent, meaning thereby that from one of these properties one may deduce the other.

Entries in column 4 of Table 2 show that the equivalence $p \equiv q$ *is true only when p and q have like truth values,* i.e., when $p \to q$ and its converse $q \to p$ are both true.

Note the equivalence of statements A, B, C of Table 3:

$$A\colon p \to q \qquad B\colon \sim q \to \sim p \qquad C\colon \sim p \vee q$$

and the important *transitive property of* equivalence:

If $A \equiv B$ and $B \equiv C$, then $A \equiv C$.

For example, if A is the statement:

If I am working, I am happy.

then B and C are respectively the equivalent statements:

If I am not happy, I am not working.
Either I am not working, or I am happy.

and the power of good symbolism has been demonstrated.

		A	B	C
p	q	$p \to q$	$\sim q \to \sim p$	$\sim p \vee q$
T	T	T	T	T
T	F	F	F	F
F	T	T	T	T
F	F	T	T	T

TABLE 3

Like the conditional, the equivalence terminology is used when p and q are totally unrelated, as in the statement:

$$[(1 + 1) = 2] \equiv [(2 + 2) = 4]$$

The statement has the truth value (T) since $1 + 1 = 2$ and $2 + 2 = 4$ are both true. Again, the statement:

$$[(1 + 1) = 3] \equiv [(2 + 2) = 5]$$

is *true* since both p: $1 + 1 = 3$ and q: $2 + 2 = 5$ are *false*.

On the other hand, the statement:

The sun is shining is equivalent to $1 + 1 = 3$.

cannot be said to be true or false until one knows the state of the weather on the day on which the statement is made. The statement is true if it is made on a day when the sun is not shining, for then statements p: "The sun is shining" and q: $1 + 1 = 3$ are both false.

Some equivalences for future use are given in Table 4. They may be established either by the transitive property of equivalence or by means of truth tables.

$$
\begin{aligned}
\sim[p \wedge q] &\equiv \sim p \vee \sim q && (1)\\
\sim[p \vee q] &\equiv \sim p \wedge \sim q && (2)\\
\sim[p \rightarrow q] &\equiv p \wedge \sim q && (3)\\
[p \rightarrow q] &\equiv \sim q \rightarrow \sim p && (4)\\
&\equiv (p \wedge \sim q) \rightarrow \sim p && (5)\\
&\equiv (p \wedge \sim q) \rightarrow q && (6)\\
&\equiv (p \wedge \sim q) \rightarrow (r \wedge \sim r) && (7)
\end{aligned}
$$

TABLE 4 *Some Useful Equivalences*

Remark The first three equivalences of Table 4 are useful in forming the denials of conjunctions, disjunctions, and conditionals. From (1) the denial of the conjunction $p \wedge q$ is the disjunction of the denials of p and q, in symbols $\sim p \vee \sim q$; from (2) the denial of the disjunction $p \vee q$ is the conjunction of the denials of p and q, in symbols $\sim p \wedge \sim q$; from (3) the denial of the conditional $p \rightarrow q$ is the conjunction $p \wedge \sim q$.

For example, from (3) it is seen that the denial of the statement:

If I am working, then I am happy.

is the statement:

I am working, and I am not happy.

Exercise 1.2

1 Write in symbolic form the contrapositive of the converse of $p \to q$.

2 State the denial of

a. the conjunction: "$1 + 1 = 2$ and $2 \times 3 = 6$."

b. the disjunction: "$1 + 1 \neq 2$ or $2 \times 3 \neq 6$."

3 a. State the converse and the contrapositive of the theorem: "If a quadrilateral is a parallelogram, its diagonals bisect each other."

b. State the denial of the theorem, its converse, and the contrapositive of this denial.

4 Show that $\sim(\sim p) \equiv p$.

5 A compound statement which is true regardless of the truth values of its components is called a *tautology*. Show that $p \to (\sim p \to q)$ is a tautology.

6 Determine by means of truth tables which of the following are tautologies:

a. $\sim(p \wedge \sim p)$

b. $[(p \vee q) \wedge \sim p] \to q$

c. $(p \wedge q) \vee [(\sim p) \wedge (\sim q)]$

d. $[(p \to q) \wedge p] \to q$

7 If p, q are the statements:

p: L_1 is parallel to L_2
q: L_2 is perpendicular to L_3

state in words the equivalent statements

a. $p \to q$

b. $\sim q \to \sim p$

c. $\sim p \vee q$

8 Select a theorem of elementary geometry of the form $p \to q$ and restate in the equivalent forms $\sim q \to \sim p$ and $\sim p \vee q$.

9 By means of truth tables, establish the equivalence

$$[p \to (q \to r)] \equiv [(p \to q) \to r]$$

a. Is the statement on the left the converse of the statement on the right? *Hint:* See ([43] pp. 71–73; [104] pp. 44–46).

b. Write a converse of each statement.

c. Determine by means of a truth table whether your converses are equivalent.

10 If a statement involves three components, p, q, r, how many rows will its truth table contain? Prepare a truth table for the conditional: "If $p \to q$ and $q \to r$, then $p \to r$."

11 Establish the equivalences of Table 4 by means of truth tables.

12 Verify that an interchange of the symbols $\vee$ and $\wedge$ in (1) of Table 4 gives (2) of the same table. When such is the case, the two equivalences are called *duals* of each other.

13 State in words the denial of each of the following:

a. A complex number is real or nonreal.

b. If $\triangle ABC$ is equilateral, it is isosceles and equiangular.

c. In $\triangle ABC$, $\angle A = \angle B$, and $\angle C = 60°$.

d. A $\triangle ABC$ has two equal angles, or the triangle is not isosceles.

1.5 Necessary and sufficient conditions

The conditional statement:

if p, then q

is often stated in the alternate forms:

q is a *necessary* condition for p.

or

p is a *sufficient* condition for q.

Consequently, if a statement "if p, then q" and its converse "if q then p" are both true, either condition $p(q)$ is both *necessary and sufficient* for the other $q(p)$. When such is the case, p is said to be equivalent to q, in symbols $p \equiv q$. A substitute terminology for $p \equiv q$ is:

p if and only if q

or

q if and only if p

Some theorems of Euclidean geometry restated in the new terminology are:

Theorem 1 A necessary *condition that a triangle be equilateral is that it be isosceles.*

Theorem 2 A sufficient *condition that a triangle be isosceles is that it be equilateral.*

Theorem 3 A necessary and sufficient *condition that a rectangle be a square is that all its sides be equal.*

These three theorems bring out the fact that a *necessary* condition may be too little, a *sufficient* condition too much, and a *necessary and sufficient* condition neither too little nor too much. Note is made, also, that a definition always contains a *necessary and sufficient* condition. Compare, for example, Theorem 3 with the dictionary definition for a square: "A four-sided plane, rectilineal figure having all its sides equal."

Exercise 1.3

1 Using the "sufficient language," restate the theorem: "If two triangles are congruent, their corresponding altitudes are equal."

2 Using the "necessary language," restate the theorem: "If a triangle is inscribed in a semicircle, it is a right triangle."

3 Is the condition that an interior point P of a triangle be equidistant from its three sides a necessary or a sufficient condition that P be the point of intersection of the angle bisectors? Why?

4 Is the condition $a/b = c/d$, $b \neq 0$, $d \neq 0$ a necessary or sufficient condition that $ad = bc$? Why?

5 Is the condition $x = 0$ a necessary or sufficient condition that

$$x^2 - 2x = 0$$

Why?

6 Give a necessary and sufficient condition that a number be equal to its negative.

7 Is the perpendicularity of its diagonals a necessary or sufficient condition that a quadrilateral be a rhombus? Why?

8 If A, B, C, D are four points of the Euclidean plane, and if lines AB and CD intersect at O, is the condition that $OA \cdot OB = OC \cdot OD$ a necessary, or sufficient, or both a necessary and a sufficient condition that the four points lie on a circle? Why?

9 Compare Playfair's Axiom of Parallels (Appendix B, Sec. 5): "There exists only one line parallel to a given line through a given point not on the line" with Euclid's parallel axiom (Appendix A, Sec. 2, Postulate 5). Are the two equivalent? Why? *Hint:* See ([96], pp. 25–27).

1.6 Formal deductive reasoning. Valid argument forms

Much of what has gone before has been a preparation for one of the main objectives of this chapter—showing when the deductive reasoning process is *correct* and when it is *incorrect.* Classical logic, a more verbal system than the one under investigation, contains definite rules for correct deductive reasoning. First of all, there are the fundamental rules of inference:

Rule 1 *If you accept the truth of the conditional "if p, then q" and the truth of the antecedent p, you must accept the truth of the consequent q.*

Rule 2 *If you accept the truth of the conditional "if p, then q," but deny the truth of the consequent q, you must deny the truth of the antecedent p.*

These rules are applied in the following examples:

Example 1

If it is raining, John is studying. (1)
It is raining. (2)
Therefore, John is studying. (3)

Example 2

If it is snowing, the temperature is below zero. (1)
The temperature is not below zero. (2)
Therefore, it is not snowing. (3)

In each example, the truth (T) of statement (3) was deduced from statements (1) and (2) assumed to be true (T). The entire process is called *deductive reasoning.*

There are other ways of describing the deductive process. The conclusion (3) is said to be the inescapable consequence of hypotheses (1) and (2); or, preferably, (3) is said to be a (logically) *true* conclusion reached by a *valid* argument. Whether the conclusion is factually true or false is immaterial. Many valid arguments may be given in which the conclusion is factually true in some instances and false in others, as the next example shows.

Example 3

If you are a member of this class, you are over twenty-one years old.
You are a member of this class.
Therefore, you are over twenty-one years old.

Even though you are actually seventeen years old, you are here bound to accept the conclusion as true, if you are a member of this class and have agreed to the original statements. There is no turning back. But this should not be a matter for concern. You have not agreed to accept factual truth but rather what must be true if certain other things are true. The point in question is brought out by the abstract argument:

If you are an X, you are a Y.

You are an X.

Therefore, you are a Y.

What is an X? A Y? These questions are immaterial. As Bertrand Russell once said, somewhat facetiously: "Mathematics is the subject in which we never know what we are talking about nor whether what we say is true." His remark makes sense when one notes that terms used in deductive reasoning are undefined, and conclusions reached rest ultimately on unproved, and sometimes meaningless, statements. In short, conclusions obtained by deductive reasoning are, therefore, independent of the nature of the elements involved and are completely detached from opinions, beliefs, facts, feelings, or emotions in any way connected with these elements.

In classical logic, special names are given to the two types of arguments of Examples 1 and 2. They are, respectively, the modus ponens and the modus tollens arguments, and symbolic logic represents them in the forms shown below.

Argument 1 (Modus Ponens)

$$\begin{array}{l} p \to q \\ \underline{p \quad\quad} \\ \therefore q \end{array}$$

Argument 2 (Modus Tollens)

$$\begin{array}{l} p \to q \\ \underline{\quad \sim q} \\ \therefore \quad \sim p \end{array}$$

The horizontal line is used in each case to separate the hypotheses from the conclusion. Although rules 1 and 2 establish the validity of these arguments, their validity may be checked also by Table 2. Argument 1 is checked first. The first two columns and column 3 of the truth table show that the only row in which the hypotheses are both true is the first, and in this row the conclusion q of the argument is true. The same table shows only one row in which the hypotheses of argument 2 are both true, and in this row $\sim p$ is true (that is, p is false). Hence, both arguments are valid.

Some other examples of valid arguments are instructive.

Argument 3 (Disjunctive Syllogism)

$$\begin{array}{l} p \vee q \\ \underline{\sim p \quad\quad} \\ \therefore q \end{array}$$

Again, Table 2, column 2, is used to check validity. There is only one row of the table in which the hypotheses are both true, and in this row the conclusion q is true.

This type of argument is used when one reasons thus: "Either AB is parallel to EF, or CD is parallel to EF. But AB is not parallel to EF, and so CD is parallel to EF."

The validity of the next argument is established by means of a truth table. (See Exercise 1.2, question 10.)

Argument 4 (Hypothetical Syllogism)

$$\begin{array}{l} p \rightarrow q \\ \underline{q \rightarrow r} \\ \therefore p \rightarrow r \end{array}$$

Such an argument is used repeatedly in solving originals, when one reasons that a theorem of the form $p \rightarrow r$ will be proved if simpler theorems $p \rightarrow q$ and $p \rightarrow r$ are proved in advance.

Checking validity by truth tables can be extremely laborious in more complicated arguments, such as argument 5. In such a case, successive units of deduction may often be used to advantage.

Argument 5

$$\begin{array}{l} p \rightarrow q \\ q \rightarrow (r \vee s) \\ \underline{p \wedge \sim r} \\ \therefore s \end{array}$$

Four units in deducing conclusion s from the hypotheses are as follows:

1 $p \rightarrow q$; $q \rightarrow (r \vee s)$; $\therefore p \rightarrow (r \vee s)$ (argument 4, hypothetical syllogism)
2 $p \wedge \sim r$; $\therefore p$ and $\sim r$ (Table 2)
3 $p \rightarrow (r \vee s)$; p; $\therefore (r \vee s)$ (argument 1, modus ponens)
4 $(r \vee s)$; $\sim r$; $\therefore s$ (argument 3, disjunctive syllogism)

and the desired conclusion s has been reached.

1.7 Invalid arguments. Incorrect reasoning

An argument in which the conclusion need not be true when the hypotheses are true is said to be *invalid*, or in less formal language, to represent incorrect reasoning. Two simple invalid arguments are obtained from the two valid arguments of Sec. 1.6, by simply interchanging the second and third statements in each.

Invalid Argument 1
(The fallacy of affirming the consequent)

$$\begin{array}{l} p \rightarrow q \\ \underline{q \qquad} \\ \therefore p \end{array}$$

Invalid Argument 2
(The fallacy of denying the antecedent)

$$\begin{array}{l} p \rightarrow q \\ \underline{\sim p \qquad} \\ \therefore \sim q \end{array}$$

Table 2, column 3, shows *two* rows, the first and third in which the hypotheses of invalid argument 1 are both true; but in the first row p *is true*, and in the third row p *is false.*

The same table shows *two* rows, the third and fourth, in which the hypotheses of invalid argument 2 are true; but in one of them the conclusion $\sim q$ is true, and in the other it is false. In neither argument, therefore, is one forced to accept the *truth* of the conclusion from the truth of the hypotheses, and this is why the arguments are invalid.

Two simple examples of faulty reasoning in elementary geometry will show how easily errors can creep into an argument.

Example 1 (Incorrect Reasoning)

If a triangle ABC is equilateral, the triangle is isosceles. (1)
Triangle ABC is not equilateral. (2)
Therefore, triangle ABC is not isosceles. (3)

The nonequilateral triangle ABC (Fig. 1.1) with $AC = BC \neq AB$ shows graphically why this argument is invalid.

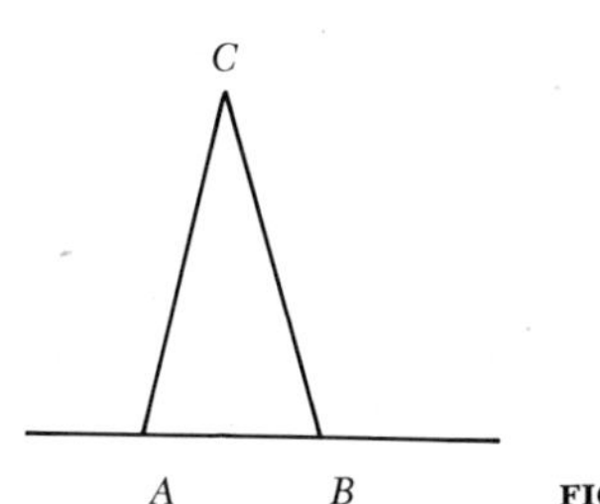

FIG. 1.1

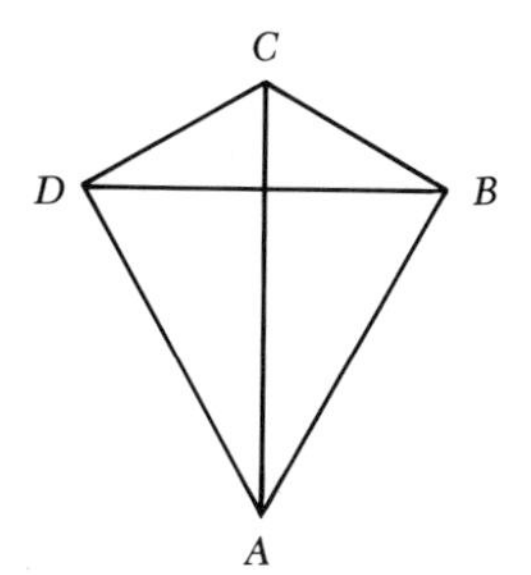

FIG. 1.2

Example 2 (Incorrect Reasoning)

If a quadrilateral is a rhombus, its diagonals are perpendicular.
The diagonals of quadrilateral $ABCD$ are perpendicular.
Therefore, quadrilateral $ABCD$ is a rhombus.

The kite-shaped quadrilateral $ABCD$ (Fig. 1.2) shows that the quadrilateral may have its diagonals perpendicular and still not be a rhombus.

The next argument is slightly more complicated.

Invalid Argument 3

$$\begin{array}{l} \sim(\sim p \wedge \sim q) \\ \underline{\sim(\sim p)\qquad\quad} \\ \therefore\ \ q \end{array}$$

From Table 4, equivalence 1, and Exercise 1.2, question 4, the argument may be written in the form

$$\begin{array}{l} p \vee q \\ \underline{p\qquad} \\ \therefore\ q \end{array}$$

That the argument is invalid is shown in Table 2, column 2. There are two rows in which the hypotheses are true; but in one of them q is true, and in the other q is false. It need not be concluded, therefore, that q is true.

Other examples of invalid arguments will be found in Exercise 1.4.

Exercise 1.4

1 Check the validity of each of the following arguments:

a. $$\begin{array}{l} p \to q \\ \underline{\sim p\qquad} \\ \therefore\ \sim q \end{array}$$

b. $$\begin{array}{l} \sim p \to \sim q \\ \underline{\sim(p \wedge \sim r)} \\ \therefore\ \sim q \end{array}$$

2 Fill in the blank in each of the arguments:

a. If two angles and the included side of one triangle equal two angles and the included side of another, the two triangles are congruent. (1)
In triangle ABC and $A'B'C'$, $\angle A = \angle A'$, $\angle B = \angle B'$, and the included sides c and c' are equal. (2)
Therefore, ____. (3)

b. If two angles of a triangle are equal, the sides opposite these angles are equal. (1)
In triangle ABC, side $a \neq$ side b. (2)
Therefore, ____. (3)

3 Check the validity of the argument whose three hypotheses are

$$(1)\ s;\ (2)\ q;\ (3)\ [p \to (q \to \sim r)] \wedge [\sim p \to \sim s]$$

and whose conclusion is $\sim r$.

1.8 The notion of proof. Direct and indirect proofs

In modern texts on logic, the fact that a statement C follows from given premises $P_1, P_2, \ldots, P_n$ is written, using the turnstile symbol:

$$P_1, P_2, \ldots, P_n \vdash C$$

and is read:

$$P_1, P_2, \ldots, P_n \text{ yields } C$$

Thus the hypotheses P_1, P_2, P_3 of argument 5, Sec. 1.6, yield the conclusion s. A proof that such is the case consists of the four units of deduction used in obtaining the desired conclusion and the proof in this case is *direct*, in accordance with the definition:

Any valid argument or sequence of valid arguments starting with the truth of initial statements $P_1, P_2, \ldots, P_n$ and ending with a conclusion C is called a direct proof that $P_1, P_2, \ldots, P_n$ yield C. Indirect proofs are often preferable to direct ones.

In an indirect proof, the investigator employs the logical principle of the excluded middle: "A statement or its denial is true." If one is disproved, the other follows.

For example, to prove that in triangle ABC with $\angle A = \angle B$ the opposite sides a and b are equal, one assumes the contrary $a \neq b$ and deduces a contradiction. From this contradiction it follows that the hypothesis $a \neq b$ is false, and hence its denial $a = b$ is true.

Again, to show that the hypotheses

$$C \to (B \wedge P) \qquad \text{and} \qquad C \to \sim B$$

yield $\sim C$, one assumes the contrary, C, and deduces a contradiction as follows: "From $C \to (B \wedge P)$ and C follows $B \wedge P$ and hence B; also, from $C \to \sim B$ and C follows $\sim B$ and a contradiction has been reached. Hence C is false and its denial $\sim C$ is true."

Symbolic logic brings into focus other methods of proof based on the equivalences 4, 5, 6, 7 of Table 4. For convenience they are repeated in Table 5 below.

$$
\begin{aligned}
(p \to q) &\leftrightarrow \sim q \to \sim p && (1)\\
&\leftrightarrow (p \wedge \sim q) \to \sim p && (2)\\
&\leftrightarrow (p \wedge \sim q) \to q && (3)\\
&\leftrightarrow (p \wedge \sim q) \to (r \wedge \sim r) && (4)
\end{aligned}
$$

TABLE 5

Each of the statements to the right of the equivalence sign shows the substitute theorem whose proof constitutes proof of the theorem "if p, then q."

In (1), the contrapositive theorem $\sim q \rightarrow \sim p$ is proved. In each of the remaining cases, the conditional on the right of the equivalence sign has for its hypothesis p and the denial of q. But in (2) the consequent is $\sim p$; in (3) it is q; and in (4) it is $r \wedge \sim r$, the familiar reductio ad absurdum of high school days.

In passing, it is noted that although indirect proofs may usually be replaced by direct ones, there are some theorems which by their very nature preclude such a possibility. See ([31] pp. 86–87).

Another fact is noted: in case a theorem resists proof, it may happen that the theorem is not true (valid). In such a case, a single counterexample will disprove the theorem. For example, the statement: "The square of every odd number is even" is disproved by showing a single odd number, such as 3, whose square is odd.

Exercise 1.5

1 a. Using the sequence of theorems in Appendix A, give an indirect proof of the theorem: "If two parallel lines are cut by a transversal, the alternate interior angles are equal." *Hint:* See ([117] pp. 70–72).

b. Which of the equivalences of Table 5 did you use in your proof?

2 Give an indirect proof of the theorem: "If the bisectors of two interior angles of a triangle are equal, the triangle is isosceles." *Hint:* See (*Am. Math. Monthly*, vol. 70, p. 79, January, 1963).

3 If P and Q are trisection points of chord AB of a circle with center O (Fig. 1.3), do lines OP and OQ trisect angle AOB? *Hint:* Prove $\angle AOP = \angle QOB = x$; then let $\angle POQ = y$ and show that $x = y$ leads to a contradiction.

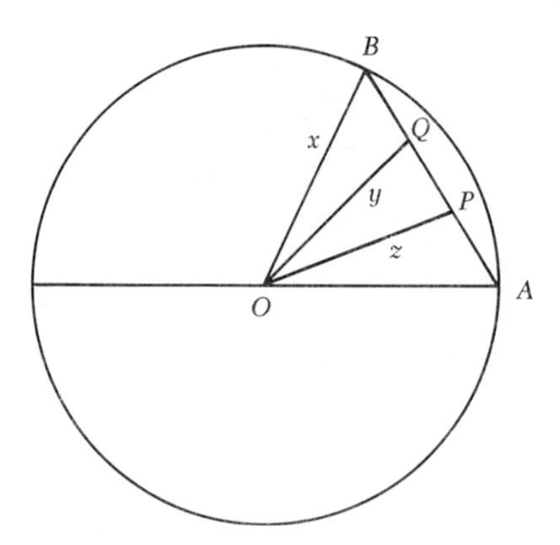

FIG. 1.3

MISCELLANEOUS EXERCISES

Exercise 1.6

1 Suppose that on a true, false examination consisting of five questions, you are given the information:

There are more true than false questions.
No three consecutive questions have the same answer.
The first and last questions have opposite answers.

Question 2 of the examination is the only one whose truth (truth value) you actually know, and this assures you of having all the answers correct. What are the correct truth values of all five questions? Give reasons for your answer. *Hint:* Does the arrangement T, F, T, F, T represent a possible set of truth values for questions 1 to 5? Why? Write all possible arrangements of the letters T, F satisfying the hypotheses and see how many are ruled out by the hypotheses.

2 Of the three prisoners A, B, C in a certain jail, A had normal vision, B had only one eye, and C was totally blind. The jailer told them that from three white hats and two red ones he would select three and put a hat on each prisoner's head. Each was prevented from seeing what color had been placed on his head. When brought together, the jailer offered freedom to prisoner A if he could tell what color was on his head. When A failed, the same privilege was extended to B. He also failed. The blind prisoner then smiled broadly and said:

I do not need to have my sight.
From what my friends with eyes have said
I clearly know my hat is ______________.

What color of hat was the blind man wearing? Give reasons for your answer.

3 Suppose that it is true that when I drink coffee, and then count sheep, I do not fall asleep; and also, that if I do not drink coffee, I am not nervous. But, I am nervous, and I count sheep. Is it true that I fall asleep? Justify your answer by exhibiting in symbolic form a valid argument whose conclusion is the answer to the question. *Hint:* See Exercise 1.4, question 3, and ([3] p. 11).

4 Consult reference [31], pp. 81–82, 86–87, and then answer:
   a. What is meant by a constructive proof?
   b. Comment on the statement: "There is an essential difference between proving the existence of an object of a certain type by constructing a tangible example of such an object and showing that if none existed one could deduce contradictory results."

1.9 Quantifiers and sets

There are many simple forms of inference which are clearly valid but whose validity cannot be checked by the preceding methods. Some initial illustrations will show how different they are from ones already considered.

Example 1

All equilateral triangles are isosceles.
All isosceles triangles have two equal sides.
Therefore, *all* equilateral triangles have two sides.

Example 2

All equilateral triangles are equiangular.
No equiangular triangle is a right triangle.
Therefore, *no* equilateral triangle is a right triangle.

Example 3

Some isosceles triangles are equilateral.
All equilateral triangles are equiangular.
Therefore, *some* isosceles triangles are equiangular.

Example 4

Things equal to the same thing are equal to each other.
$\angle a + \angle b$ and $\angle b + \angle c$ are straight angles.
Therefore, $\angle a + \angle b = \angle b + \angle c$.

Even a superficial examination of these arguments will show that they differ from those already discussed in two important respects: (1) they involve the notion of quantity expressed by such words as "all," "some," "no," and their synonyms; (2) they take into account the internal structure and meaning of statements, unlike the logic just discussed where statements were unanalyzed.

In the literature, words involving quantity are called "quantifiers," statements involving them are quantified, and that branch of logic which discusses their role is the theory of quantification. It is as old as Euclid. The modern treatment of quantified statements involves the notion of a set, class, or collection. Set language and theory have been in high fashion in mathematics ever since George Cantor (1845–1918) gave his definition of the word "set" and mathematicians were immediately alerted to the difficulties involved in an unrestricted theory of sets.

Sets and figures representing them appear here in descriptions of such important notions as *relations, functions,* and *operations.*

1.10 Set terminology, symbolism, operations

The word "set" is used here in the colloquial sense: Elements or objects x with a given property form a set A.

The fact that x does (does not) belong to A is written symbolically:

$$x \in A (x \notin A)$$

Most familiar sets, such as the days of the week, the integers, or the students in the classroom, have the property that the set is not a member of itself. Such a set is called an *ordinary* set. A set S of students is an ordinary set, since S is not a student and hence is not a member of itself. Only ordinary sets are considered here.

In what follows, the letter I will denote a fixed or universal set; and A, B, C, . . . subsets of I where by a subset A of a set B is meant a set every element of which belongs to B. The symbol $A \subseteq B$ (or $B \supseteq A$) is read:

A is a subset of B, or B includes A.

or in less formal terms, "All A is B." A is said to be a *proper* subset of B, in symbols $A \subset B$, if and only if every element of A is an element of B but not every element of B is an element of A.

For visualizing and grasping this definition and others of set theory, Venn diagrams will be helpful. In such a diagram, the universal set I is represented by points within a rectangle (Fig. 1.4), and subsets A, B, . . . by points inside circles lying within the rectangle. Such diagrams often suggest methods by which statements about sets can be proved or disproved, but they are in no sense substitutes for formal proofs.

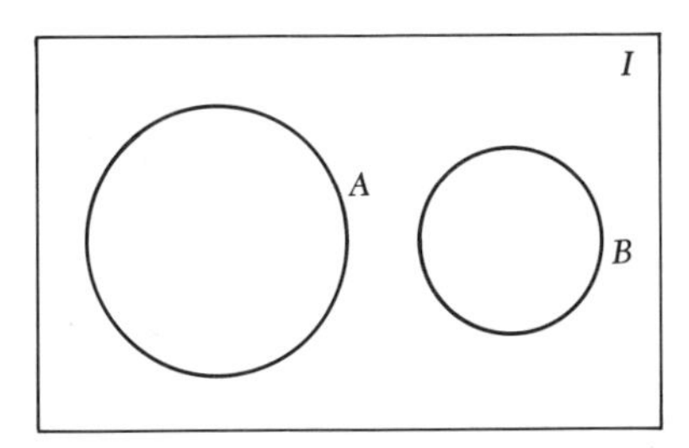

FIG. 1.4

However paradoxical it may seem, there is a null set, denoted by $\varnothing$, by which is meant *a set without any elements. By convention, the null set is a subset of every set.*

The set of coins in your purse after you have spent all your money is an example of a null set. Another one is the set of all women presidents in the United States in the year 1964. (It may not be a good example in the year 2000!)

The existence symbol $\exists$ is used to show that a set is not empty. Thus

$$\exists x \qquad \ni x \in A$$

is read: "There exists an x such that x belongs to A."

If both $A \subset B$ and $B \subset A$, then clearly every element of one set belongs to the other and the two sets are equal, as indicated by the symbol: $A = B$.

With the appearance of the null set and the equality symbol, a mathematical theory of sets begins to emerge. The theory is based on the fact that sets may be combined to form other sets—just as in arithmetic, numbers may be added or multiplied to form other numbers.

Analogous to the familiar operations of addition and multiplication of numbers are the operations on sets called *union* and *intersection* (also called, respectively, logical sum and logical product). They are denoted by the respective symbols $\cup$ and $\cap$ sometimes called "cup" and "cap."

Definition 1 *By the union $A \cup B$ of two subsets A and B of I is meant the set of elements x such that*

$$x \in A \qquad \text{or} \qquad x \in B$$

as shown by points of the shaded region of Fig. 1.5. ("Or" is used here in the nonexclusive sense, that is, x may be common to both A and B.)

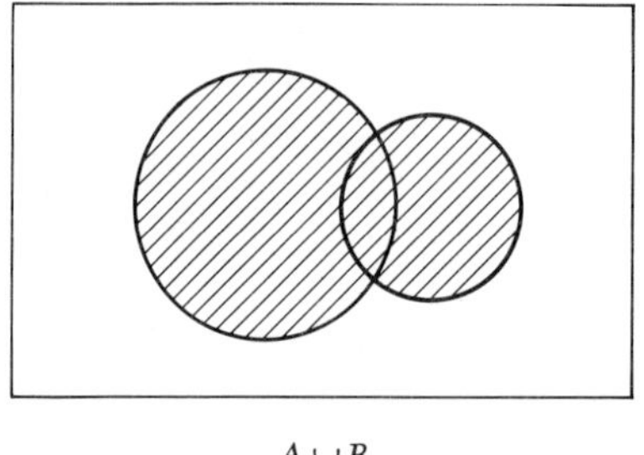

$A \cup B$

FIG. 1.5

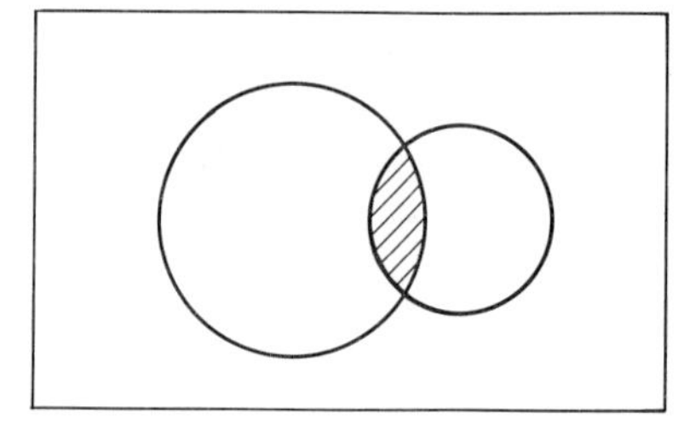

$A \cap B$

FIG. 1.6

Definition 2 *By the intersection $A \cap B$ of two subsets A and B of I is meant the set of elements x such that*

$$x \in A \qquad \text{and} \qquad x \in B$$

as shown by points of the shaded region of Fig. 1.6.

Special interest is attached later to the product set $A \cap B$ when this set is A itself and $A \cap B = A$ (Fig. 1.7*a*), or is the null set and $A \cap B = \varnothing$ (Fig. 1.7*b*), or is not the null set and $A \cap B \neq \varnothing$ (Fig. 1.7*c*).

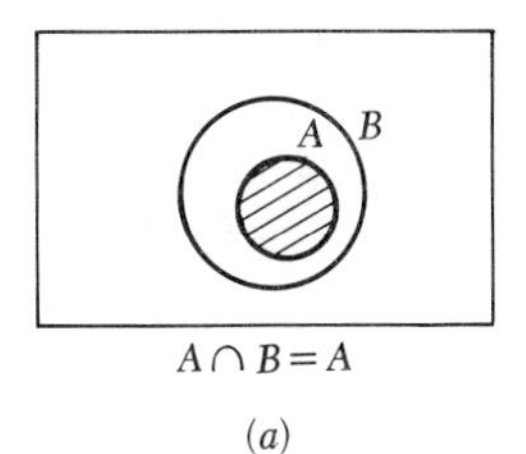

$A \cap B = A$

(*a*)

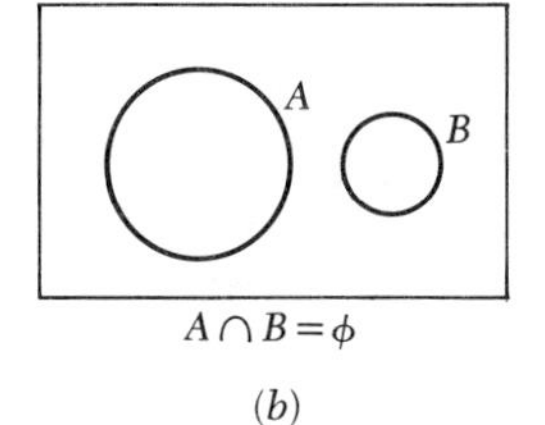

$A \cap B = \phi$

(*b*)

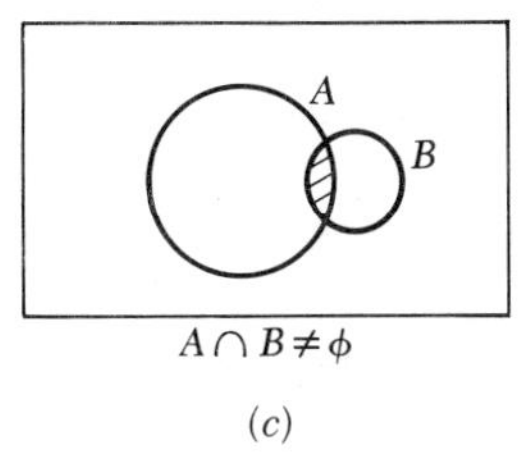

$A \cap B \neq \phi$

(*c*)

FIG. 1.7

For convenience in reference the customary terminology associated with these various cases is tabulated in Table 6.

$A \cap B = A$	All A is B	Fig. 1.7*a*
$A \cap B = \varnothing$	No A is B	Fig. 1.7*b*
$A \cap B \neq \varnothing$	Some A's are B's	Fig. 1.7*c*

TABLE 6

There are many examples in geometry of the union and intersection of sets. If A is the set of points on a line not in a plane p, and B the sets of points in the plane, the union set $A \cup B$ consists of points of the line together with points of the plane. The product set $A \cap B$ is the null set if the line is parallel to the plane and is the single point of intersection of the line and the plane when they intersect.

Definition 3 *By the complement A' of subset A of I is meant those elements of I not belonging to A.*

The complement A' of A is represented by points of the shaded region of Fig. 1.8.

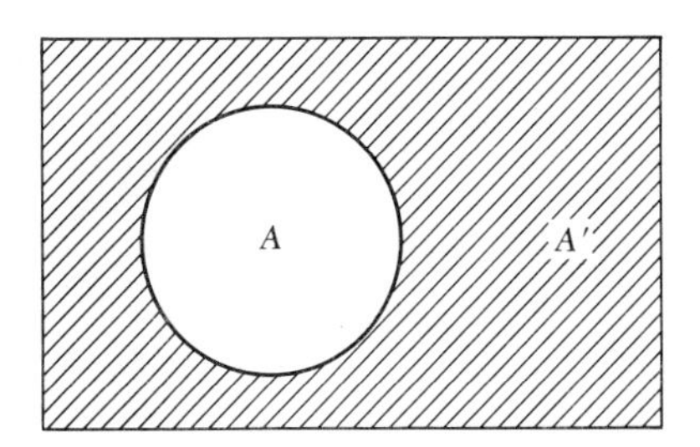

FIG. 1.8

If, for example, L is the set of points on a line of the plane I, the complement L' of L will be all points of the plane not lying on the line.

1.11 A circle test for validity

A graphic method for testing the validity of arguments involving quantifiers makes use of Venn diagrams and Table 6. The method called the *circle test* is described for some classical syllogistic arguments, the first of which involves only the quantifier "all."

Argument 1

All A's are B's.
All B's are C's.
Therefore, all A's are C's.

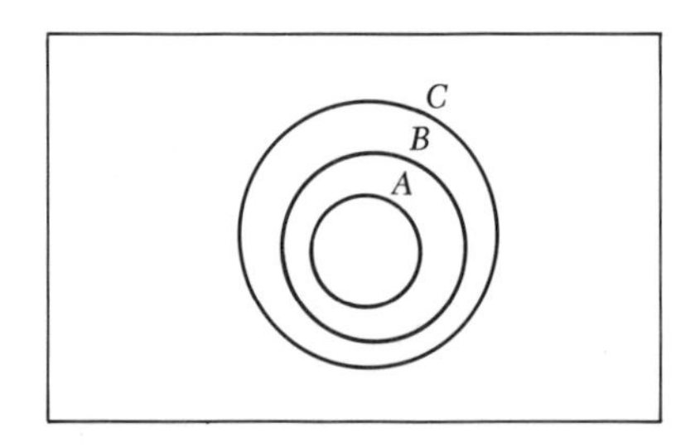

FIG. 1.9

Let elements of sets A, B, and C be represented by points of circles labeled A, B, C (Fig. 1.9). From the first hypothesis, circle A lies within circle B; and from the second, circle B lies within circle C. Circle A therefore lies within circle C, or in set terminology: "All A's are C's," and the validity of argument 1 has been checked. It is illustrated for particular sets A, B, C by Example 1 of Sec. 1.9.

The next argument, illustrated by Example 2 of Sec. 1.9, involves the quantifier "no."

Argument 2

All A's are B's.
No B's are C's.
Therefore, no A is C.

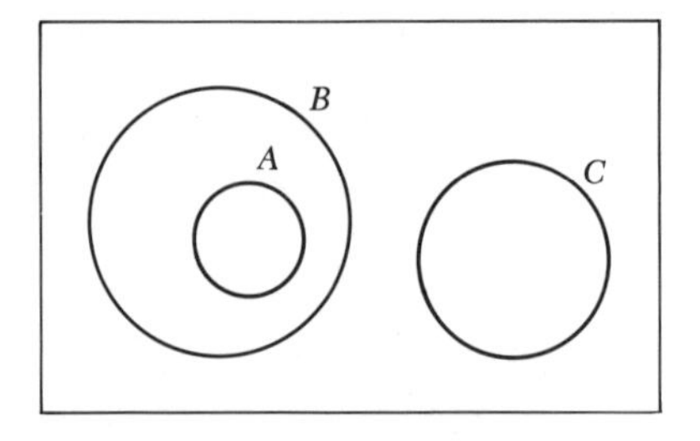

FIG. 1.10

From the first hypothesis, circle A is within circle B (Fig. 1.10); and from the second, circle B is without circle C. Circle A is therefore without circle C, or in set terminology: "No A is C," and the argument is valid.

The third argument, illustrated by Example 3 of Sec. 1.9, involves the quantifier "some."

Argument 3

Some A's are B's.

All B's are C's.

Therefore, some A's are C's.

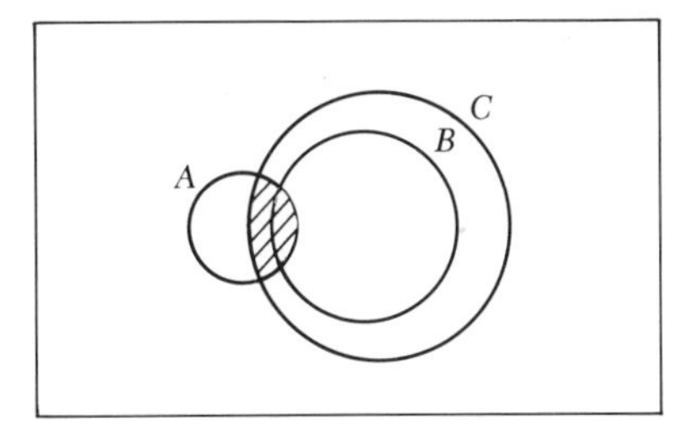

FIG. 1.11

From the first hypothesis, circles A and B intersect in points of the shaded region of Fig. 1.11; and from the second, circle B lies within circle C. Therefore, points of A which are in B are also in C, or in set terminology: "Some A's are C's," and argument 3 is valid.

Example 4 of Sec. 1.9 is used now in a partial proof of the elementary theorem: "If two straight lines intersect, the vertical angles are equal."

If $\angle a$ and $\angle c$ are any two vertical angles formed by the lines AB and CD intersecting at O (Fig. 1.12) and if b is their common supplement, the conclusion from the example cited above is

$$\angle(a + b) = \angle(b + c)$$

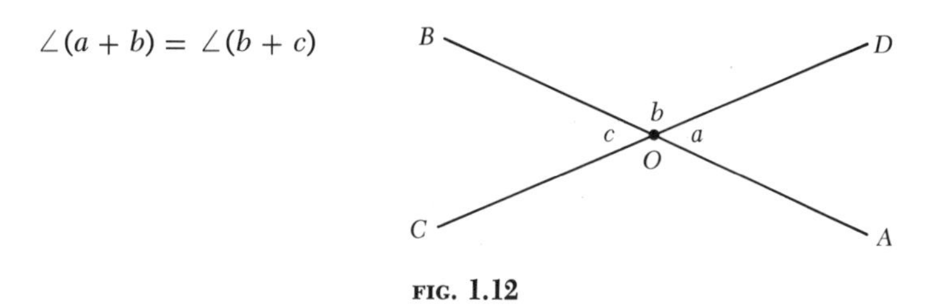

FIG. 1.12

The argument is checked graphically. Let circle A (Fig. 1.13) represent the set of things equal to each other; circle B, the set equal to the same thing; and finally circle C, the pair of angles $a + b$ and $b + c$. Since each of the latter is a straight angle, circle C lies within circle B; and since things equal to the same thing are equal to each other, circle B lies within circle A. Circle

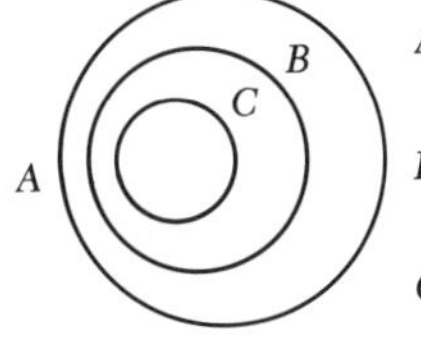

A–Things equal to each other

B–Things equal to the same thing

C–The angles $a+b$ and $b+c$

FIG. 1.13

C therefore lies within circle A, or in other words, $\angle(a + b) = \angle(b + c)$, and the conclusion is valid.

The circle test is used finally to show why an argument is invalid.

Argument 4 *(Invalid)*

> No undergraduates have B.A. degrees.
> No freshmen have B.A. degrees.
> Therefore, freshmen are undergraduates.

Let circle A (Fig. 1.14) represent people with B.A. degrees; circle B, undergraduates; and circle C, freshmen. Then from the first hypothesis, circles A and B are nonintersecting; and from the second, C and A are nonintersecting as shown in each of Figs. 1.14a to d. But no information is given about the position of circle C with respect to circle B. Circle C might lie within circle B (Fig. 1.14a), be external to B (Fig. 1.14b), overlap B (Fig. 1.14c), or include B (Fig. 1.14d). The only figure which justifies the conclusion of the argument is Fig. 1.14a, and since one is not forced to accept this and reject Figs. 1.14b to d, the argument is invalid.

The example is excellent for showing a conclusion which is factually true, but not logically true.

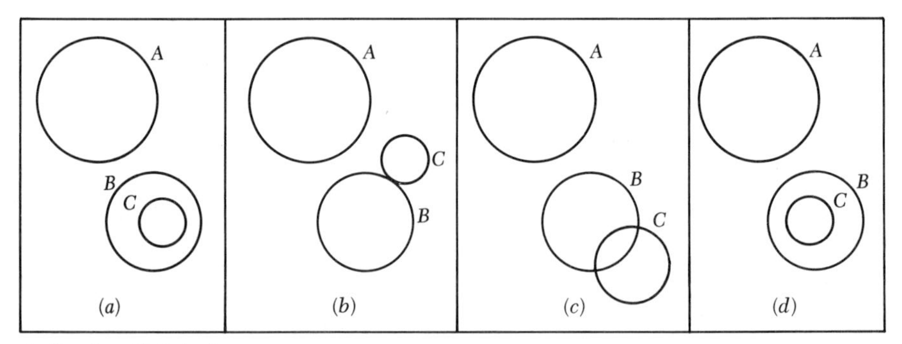

A: People with B.A. degrees
B: Undergraduates
C: Freshmen

FIG. 1.14

Exercise 1.7

Use the circle test to check the validity of each of the following:

1 No equilateral triangle is a right triangle.
All equiangular triangles are equilateral.
Therefore, no equiangular triangle is a right triangle.

2 Some parallelograms are squares.
All squares are rectangles.
Therefore, some parallelograms are rectangles.

3 No prime integers greater than 2 are even.
x is a prime integer greater than 2.
Therefore, x is not an even integer.

4 Some primes are integers divisible by 2.
All integers divisible by 2 are even.
Some primes are even integers.

5 All rectangles are parallelograms.
Not all quadrilaterals are rectangles.
Therefore, not all quadrilaterals are parallelograms.

Fill in the blank in each of the following:

6 Some negative numbers are rational.
No rational numbers are imaginary.
Therefore, ______________________________.

7 All squares are rectangles.
All rectangles are parallelograms.
Therefore, ______________________________.

8 If all lines through point P intersect line AB, then there are no parallels to AB through P.
There is a parallel to AB through P.
Therefore, ______________________________.

1.12 An algebra of sets. Reasoning by algebraic methods

From the definition of the complement A' of A follow the relations

$$A \cap A = \varnothing \qquad \text{and} \qquad A \cup A' = I$$

They suggest the algebraic relations for numbers a and a':

$$a + a' = 0 \qquad \text{and} \qquad a \cdot a'' = 1$$

in which union $\cup$ and intersection $\cap$ are replaced by addition and multiplication, and a' in the first equation is the negative of a. In the second equation a'' is the reciprocal of a.

This analogy between set *relations* and algebraic laws goes further. From the definitions of Sec. 1.10 will follow immediately the laws below in which A, B, C are sets; A', B', C', their respective complements; I, the universal set; and $\varnothing$, the null set.

SET LAWS

Intersection and union are commutative:

$$A \cup B = B \cup A \tag{1}$$
$$A \cap B = B \cap A \tag{2}$$

and associative:

$$(A \cap B) \cap C = A \cap (B \cap C) \tag{3}$$
$$(A \cup B) \cup C = A \cup (B \cup C) \tag{4}$$

There are two distributive laws:

$$A \cap (B \cup C) = (A \cap B) \cup (A \cap C) \tag{5}$$
$$A \cup (B \cap C) = (A \cup B) \cap (A \cup C) \tag{6}$$

and the sets I, $\varnothing$, and complement sets obey the laws:

$$A \cap I = A \tag{7}$$
$$A \cap A' = \varnothing \tag{8}$$
$$A \cup A' = I \tag{9}$$
$$A \cup \varnothing = A \tag{10}$$

These laws constitute the beginnings of an algebra of sets.

To see how this new algebra may be used in the reasoning process, consider the simple argument:[1]

All gentlemen are polite.
No bankers are polite.
No gentlemen are bankers.

Let g, p, b represent the respective sets: gentlemen, polite people, and bankers. Also, let the union and intersection of two sets g and p be denoted $g + p$ and gp respectively. Then, from the first hypothesis and Table 6, $gp' = \varnothing$; from the second, $bp = \varnothing$; and from the definition of the null set and the product set,

$$bgp' = \varnothing \qquad bgp = \varnothing$$

and hence

$$bgp' + bgp = \varnothing$$

[1] See ([84] p. 41).

But, as in algebra, $bgp' + bgp = bg(p + p') = bgI = bg$, and hence

$$bg = \varnothing$$

which translated says: "No gentlemen are bankers." This is a conclusion which has been reached by purely algebraic processes.

One might reasonably ask what advantage the new method has over the older methods. The answer is that the algebraic method is a more formal, rigorous approach to the reasoning process and is not subject to the errors arising from the rather haphazard method of drawing conclusions from figures.

In short, there is economy of effort in reasoning by algebraic methods. There is also clarity. And above all, there is a method for handling complicated arguments, where the limits of practical intelligibility by the older verbal and graphical methods are soon reached.

It is tempting to pursue the subject further, but to do so would require too great a digression. Learning to reason by algebraic methods takes time, patience, and practice. The present remarks simply indicate in a small way the added power of algebraic reasoning methods.

1.13 A list of reference symbols

For convenience in reference, standard symbols used in this chapter and elsewhere in the text are listed below, together with sections of the text in which they are introduced.

$\wedge$ and Sec. 1.3
$\vee$ or Sec. 1.3
$\rightarrow$ if-then Sec. 1.3
$\leftrightarrow$ if and only if Sec. 1.3
$\sim$ not (or denial) Sec. 1.3
( ) parentheses for groupings Sec. 1.3
[ ] brackets for groupings Sec. 1.3
{ } braces for groupings Sec. 1.3
$\vdash$ yields Sec. 1.8
$\ni$ such that Sec. 1.10
$\in$ belongs to Sec. 1.10
$\notin$ does not belong to Sec. 1.10
$\subset$ subset of Sec. 1.10
$\not\subset$ not a subset of Sec. 1.10
$\cup$ union Sec. 1.10
$\cap$ intersection Sec. 1.10
A' complement of A Sec. 1.10
$\exists x$ there exists an x Sec. 1.10

Exercise 1.8

1 Does the universal set I have a complement? Explain.

2 If I is the set of all positive integers, A the set of even integers, and B the set of odd integers, describe in words each of the sets (a) $A \cup B$ and (b) $A \cap B$.

3 If I is the set of real numbers and A is the set consisting of zero and the positive real numbers, describe the complement set A' to A.

4 If L_1 and L_2 are two distinct lines of the plane p, describe the sets (a) $L_1 \cup L_2$ and (b) $L_1 \cap L_2$ according as the lines L_1 and L_2 do or do not intersect.

5 If p_1 and p_2 are two distinct planes, describe each of the sets $p_1 \cup p_2$ and $p_1 \cap p_2$: (a) if the planes intersect, (b) if the planes do not intersect.

6 Using Venn diagrams, verify each of the laws:

a. $A \cap A = A$

b. $(A \cap B)' = A' \cup B'$

c. $(A \cup B)' = A' \cap B'$

A quiz on Chap. 1.

1 a. If A and B are the respective statements

$$A\colon (p \wedge q) \rightarrow r$$
$$B\colon p \rightarrow (q \rightarrow r)$$

show the equivalence of A and B by means of a truth table.

b. Given the statements

$$A\colon r \rightarrow (p \wedge q) \qquad \text{and} \qquad B\colon (q \rightarrow r) \rightarrow p$$

are the statements

$$A'\colon (r \rightarrow p) \wedge q \qquad B'\colon q \rightarrow (r \rightarrow p)$$

converses of A and B? Why?

c. Is $A' \equiv B'$? Justify your answer by means of a truth table.

2 a. Write in symbolic form the theorem and a converse of the theorem: "If the bisector of vertex angle C of triangle ABC is perpendicular to the base and bisects it, then triangle ABC is isosceles."

b. State your converse in words.

3 a. By means of a truth table check the (logical) truth of

$$\sim(p \vee q) \equiv \sim p \wedge \sim q$$

b. Write in words the denial of the statement: "I shall study or fail the course."

4 Is the following argument valid? Why?

$$\begin{array}{l} \sim(p \vee q) \\ r \rightarrow p \\ \hline \therefore\ r \end{array}$$

(*Hint:* Use equivalence of 3*a* above.)

5 Using the indirect method of proof, check the validity of the argument:

$$\begin{array}{l} C \rightarrow (B \wedge p) \\ C \rightarrow \sim B \\ \hline \therefore\ \sim C \end{array}$$

6 Write in symbolic form and check the validity of the following argument:

If $\triangle ABC$ is a right triangle, then $c^2 = a^2 + b^2$. (1)
If $c^2 = a^2 + b^2$, then $\triangle DEF$ is a right triangle *or* an equilateral triangle. (2)
$\triangle ABC$ is a right triangle *and* $\triangle DEF$ is not a right triangle. (3)
$\therefore$ $\triangle DEF$ is equilateral. (4)

7 Given the following argument:

No right triangle is equiangular.
All equiangular triangles are isosceles.
Some right triangles are isosceles.

a. Is the conclusion a theorem of Euclidean geometry?
b. Is the argument valid? Justify your answer by means of a Venn diagram.

Concluding remarks

At the end of this chapter on logic and set theory, it is well to look back at what has been done and see why it has been done.

Some preliminary remarks on inductive reasoning prefaced a more than cursory analysis of deductive reasoning by modern symbolic methods. The analysis included a study of the form, meaning, and truth of those mathematical statements in which the connectives:

and   or   if-then   if and only if

were used. Truth tables and the symbolic representation of statements were used to symbolize and mechanize the reasoning process and to distinguish correct from incorrect reasoning.

This was followed by a brief introduction to the notion of a proof, both direct and indirect. And finally set terminology, symbolism, and operations on sets were introduced, together with Venn diagrams, to aid reasoning processes in quantification theory.

One overall aim was that of replacing the somewhat vague, intuitive logic of the average person by a more rigorous, precise science—thus meeting the growing demand for more logical rigor in today's mathematics.

Other objectives were kept in mind. Continued use in this and later chapters of the new terminology and symbolism of modern mathematics familiarizes the reader with the new language and equipment he needs for advanced study.

The entire chapter is background material for everything that follows, but specifically for Chap. 2, where axiomatic systems are analyzed and illustrated and flaws in Euclid are pointed out. The study emphasizes the great changes that have taken place in geometric thinking since Euclid gave to the world its first introduction to an abstract science.

Suggestions for further reading[1]

Ambrose, Alice, and Morris Lazerowitz: "Logic: the Theory of Formal Inference."
Basson, A. H., and D. J. O'Connor: "Introduction to Symbolic Logic."
Bell, E. T.: "The Search for Truth."
Bentley, A. F.: "Linguistic Analysis of Mathematics."
Carnap, Rudolf: Foundations of Logic and Mathematics, "International Encyclopedia of Unified Science," vol. I, no. 3.
Cohen, M., and E. Nagel: "Introduction to Logic and the Scientific Method."
Copi, I. M.: "Symbolic Logic." "Elementary Mathematics of Sets," Mathematical Association of America, Committee on the Undergraduate Program.
Exner, R. M., and M. F. Rosskopf: "Logic in Elementary Mathematics."
Fujii, John N.: "An Introduction to Elements of Mathematics."
Hilbert, D., and W. Ackerman: "Principles of Mathematical Logic."
Langer, Susanne K.: "Introduction to Symbolic Logic."
Lewis, C. I., and C. H. Langford: "Symbolic Logic."
Newman, J. R.: History of Symbolic Logic, in "The World of Mathematics," vol. III.
Nidditch, P. H.: "Elementary Logic of Science and Mathematics."
Rosser, J. Barkley: "Logic for Mathematicians."
Russell, Bertrand: "Mysticism and Logic and Other Essays."
School Mathematics Study Group: "Some Basic Mathematical Concepts."
Stabler, E. R.: "An Introduction to Mathematical Thought."
Stoll, Robert R.: "Sets, Logic and Axiomatic Theories."
Suppes, Patrick: "Introduction to Logic."
Tarski, A. "An Introduction to Logic."
Weyl, Herman: Mathematics and Logic, *Am. Math. Monthly.*
Whitesitt, J. E.: "Boolean Algebra and Its Applications."

[1]For complete publication data see the Bibliography.

2 axiomatic systems

Just as it would be hard to visualize a sleek new convertible car by examining its various parts, unassembled, so too it is difficult to grasp the idea of an axiomatic system by examining its separate parts: (1) *undefined elements,* (2) *axioms,* and (3) *logical reasoning.* It is best to look first at some assembled products and then to take them apart.

Euclidean geometry is perhaps one of the most famous examples of an axiomatic system, but there are simpler ones which show the interrelation of the various parts without the added complications of this classical system.

It is proposed to examine some simple axiomatic systems first and then to follow them by others of increasing complexity. What will make the study particularly interesting will be seeing along the way the possibility of creating geometries different from Euclid's. Today, such a possibility seems quite natural, but this is a modern viewpoint which was reached originally only via a long road up. Euclid's successors have made their own valuable contributions to his beginnings, and a new chapter has been written in the book of knowledge. The following analysis of axiomatic systems reflects modern thinking as opposed to that of Euclid's day.

2.1 A simple axiomatic system and a finite geometry

To inject a note of reality into the present discussion and to avoid being too technical or abstract at the start, consider an ornament S consisting of a number of beads arranged on a set of wires in accordance with the rules:

1 Each pair of wires in S is on one and only one bead.
2 Each bead in S is on two and only two wires.
3 The total number of wires in S is four.

What further information may be deduced from these simple facts? Two somewhat trivial statements following immediately from (1) and (2) are:

4 On every pair of wires is one and only one bead.
5 Every bead is on two and only two wires.

To deduce further information, let a wire of S be denoted W_i. Then by (3), $i = 1, 2, 3, 4$; and by (2), a pair of wires W_i and W_j has a common bead, denoted simply (i,j); and since two things may be selected from four in six different ways, there are six beads in S. They are:

$A(1,2)$ $B(1,3)$ $C(1,4)$ $D(2,3)$ $E(2,4)$ $F(3,4)$

From (1) these beads are all distinct, and another deduced statement is:

6 There are exactly six beads in S.

Since also by (1) a wire, say W_1, contains a bead corresponding to each of the three remaining wires, and by (2) no two wires have more than one common bead, a seventh conclusion is:

7 There are exactly three beads on each wire.

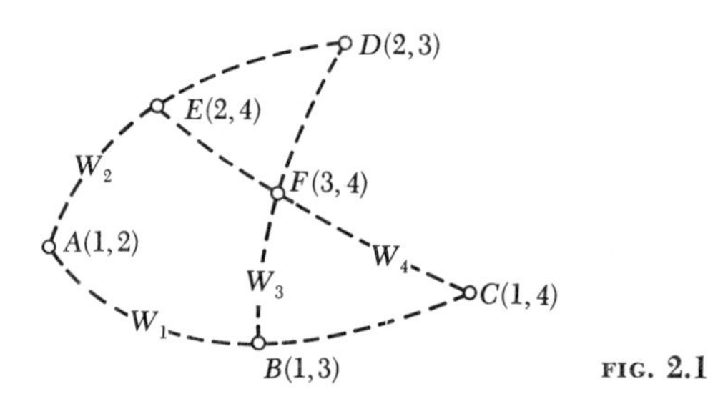

FIG. 2.1

If now beads are represented by dots, and wires by curved lines, Fig. 2.1 shows an ornament S satisfying these seven statements, and the power of deductive reasoning has been demonstrated.

Suppose now that one frees himself of physical considerations. Instead of talking about beads on a wire, let S be a set of undefined elements called points. Certain subsets of S are called lines, and all that is known at the start about these undefined elements is contained in the statements below.

AXIOMATIC SYSTEM I

Axioms 1 *Each pair of lines of S is on one and only one point.*
2 *Each point of S is on two and only two lines.*
3 *The total number of lines of S is four.*

Because the elements, point and line, with which one is dealing are undefined, the system is called abstract. Because the undefined elements are usually associated with geometric entities, the system is called a geometry.

Axioms 1 and 2 are called incidence axioms since they deal with the property of a point being on a line and a line being on a point.

Axiom 3 is an existence axiom since it guarantees the fact that the set of lines under discussion is not empty.

As for the content of the system, some results are already at hand. Since the reasoning process is independent of the nature of the elements involved, it is immaterial whether one reasons about *beads on a wire* or *points on a line*. Suitable replacements in statements 6 and 7 give, without further proof, two theorems of the abstract system:

Theorem 1 *There are exactly six points in S.*

Theorem 2 *There are exactly three points on each line of S.*

A geometry such as this one, in which a line contains only a finite number of points, is by definition a finite geometry. Also, beads and wires represent a model of this abstract geometry in accordance with the definition:

> *Any set of elements satisfying the axioms of an abstract system is said to be a model of the system.*

Another model could be obtained by replacing the words "point" and "line" by "men" and "committees." Then Axioms 1 to 3 read as follows:

> *Any two men of S are on one and only one committee.*
> *Any man of S is on two and only two committees.*
> *The total number of committees in S is four.*

Theorems 1 and 2 read:

> *There are exactly six men in S.*
> *There are exactly three men on each committee.*

The economy of the axiomatic method should now be apparent. Once an abstract system has been developed, its theorems may be used to supply information about any set of elements modeling the system. Thus, a senseless repetition of proofs is avoided. Also, if the model consists of physical objects, theorems of the abstract system often reveal physical properties of the elements so deeply hidden that they would be practically impossible to determine by observation or experimentation.

The models of this section are used next to illustrate the notion of a one to one (1-1) correspondence and an isomorphism.

2.2 One to one (1-1) correspondence. Isomorphism

The notion of a one to one correspondence is basic in practically all of mathematics and for that matter in everyday life as well. A child, for instance, learns to count by setting up a 1-1 correspondence between his fingers and the numbers 1, 2, . . . , 10.

Again, there is usually a 1-1 correspondence between the chairs and the students in a classroom. The correspondence is 1-1 when each student has a seat and each seat contains one and only one student.

If there are empty chairs or more students than chairs in the room, the correspondence ceases to be 1-1.

In the two models of Axiomatic System I (Sec. 2.1) there was a 1-1 correspondence between beads $b_1, b_2, \ldots$ and men $m_1, m_2, \ldots$ and also between wires $w_1, w_2, \ldots$ and committees $c_1, c_2, \ldots,$ where b_i correspond to m_i, and w_i to c_i. Also, in both models there was an "on" relation satisfying a set of axioms. Beads were *on* wires, and men were *on* committees. Furthermore, to each statement about a bead b_i on a wire w_i corresponded a statement about a man m_i on a committee c_i. *The two models are then said to be isomorphic with respect to the mathematical system of which they are models.*

A mathematical system is said to be *categorical* when every two models are isomorphic. In simpler language this means that when two models of a system have the same form or structure there is essentially only one representation of the mathematical system.

2.3 Undefined elements and axioms

The foregoing study of Axiomatic System I gives a first introduction to the role of undefined terms and axioms. The undefined terms are the basic elements in terms of which all other elements are to be defined.

Point and line are usually taken to be the undefined terms of elementary geometry, but there exist geometries in which the undefined elements are circles and spheres, number pairs in a plane, or even other elements depending upon the particular type of geometry in question.

To say that point and line are undefined may be puzzling in view of the fact that any standard dictionary defines them. However, a definition simply gives the meaning of a word in terms of other simpler words, or rather words whose meaning is already clear; and these simple words are in turn defined in terms of even simpler words, and so on. Such a process would therefore lead to an endless regression if it were not agreed, as is done, that certain basic words are to be left undefined.

For example, a line segment is, by definition, that portion of a line lying between two given points of the line. Here, point and line are undefined, as is also the word "between."

When Euclid defined a line as "length without breadth" and a straight line as a "line which lies evenly between two of its points," he had not defined the words "length," "breadth," and "lies evenly," and so he might just as well have taken the line itself to be an undefined term.

Just as certain elements are chosen as fundamental ones, in terms of which all others are to be defined, so too some simple statements concerning the undefined elements are chosen as fundamental, in the sense that all other statements of the system are to be deduced from them by logical reasoning.

These fundamental statements which are accepted without proof *are called axioms.* They are the foundation blocks, so to speak, upon which the structure rests.

It is proposed to study next the effect on a structure or building of a change in one or more of its foundation blocks.

2.4 A second axiomatic system

Another geometry is suggested by the familiar situation, mentioned in Sec. 2.1, of men assembling in a room S and forming committees for some particular purpose.

This time the rules are more extensive than the ones given for Axiomatic System I.

1 If A_1 and A_2 are distinct men of S, there is one and only one committee containing both of them.
2 Any two committees have at least one man in common. (This means that a man serves on at least two committees.)
3 There is at least one committee.
4 Every committee contains exactly three men of S.
5 Not all men serve on the same committee.

Unlike rules of Axiomatic System I, no mention is made of the exact number of committees in S. This information will be obtained by means of a theorem of an abstract system in which the words "men" and "committees" are replaced by the words "point" and "line." The rules above then become:

AXIOMATIC SYSTEM II

Axioms 1 *If A_1 and A_2 are distinct points of S, $\exists$ one and only one line containing both of them.*
2 *Any two lines of S intersect.*
3 *$\exists$ at least one line in S.*
4 *Every line of S contains exactly three points.*
5 *Not all points of S are on the same line.*

The number of points in S is determined by logical reasoning from these axioms. From Axioms 3 to 5, the set S contains at least four points. Let them be denoted by the numbers 1, 2, 3, 4, and let lines be denoted by triples of these numbers, or where the third point is immaterial by a number pair. Thus (1,2) denotes a line, and if 3 is a third point of the line, (1,2,3) denotes the same line.

Now three lines of S are

$$(1,2,x) \qquad (2,3,y) \qquad (1,3,z)$$

and by Axiom 1, the points x, y, z are distinct from the points 1, 2, 3 and from each other. Call them 4, 5, and 7. Then three lines of S are

$$(1,2,4) \qquad (2,3,5) \qquad (1,3,7)$$

But now by Axiom 2, lines (1,5) and (2,7) must intersect in a point P, which by Axiom 1 cannot be 3, 4, or 7. Hence S must contain a seventh point, 6, and at least the seven lines represented by the number triples in the columns of the array:

$$\begin{array}{ccccccc} 1 & 2 & 3 & 4 & 5 & 6 & 7 \\ 2 & 3 & 4 & 5 & 6 & 7 & 1 \\ 4 & 5 & 6 & 7 & 1 & 2 & 3 \end{array}$$

To show that there cannot be more than seven points in S, let 8 be another point of S. Consider the lines

$$(1,8) \qquad \text{and} \qquad (2,6,7)$$

By Axiom 2, these lines meet in a point, say Q, which cannot be the point 2, 6, or 7, for the line (1,2) already contains the third point 4; the line (1,6), the point 5; and the line (1,7), the point 3. But then Q is a fourth point on a line and Axiom 4 is contradicted. From this contradiction follows the theorem:

Theorem 1 *There are exactly seven points in S.*

A model of this axiomatic system is shown in Fig. 2.2, where a numbered dot represents a point and a triple of numbers represents a line.

The figure is a trifle misleading. A number triple such as (1,3,7) representing a line is shown on a line, but the number triple (2,3,5) representing another line is on a circle. Furthermore there are no points between two points as the figure seems to suggest. But the figure does illustrate the next theorem. (See Exercise 2.1, question 4a.)

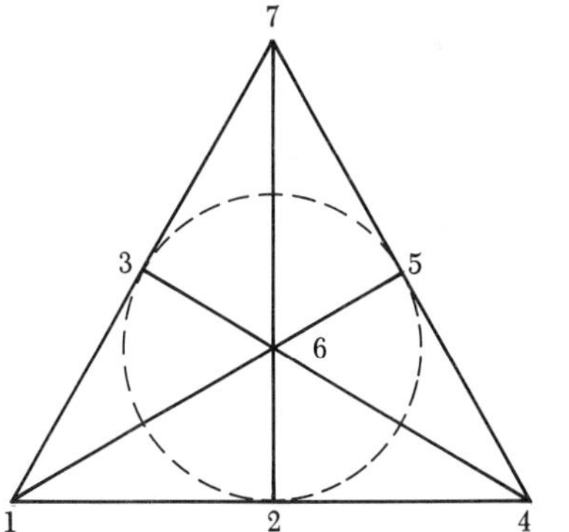

FIG. 2.2

Theorem 2 *Through each point of S pass three lines.*

For example, through point 3 (Fig. 2.2) pass the three distinct lines (3,1,7), (3,4,6), and (3,2,5).

2.5 A third axiomatic system. Affine geometry

The simple systems studied so far are incidence geometries, meaning that they deal with a point on (incident with) a line and a line on (incident with) a point. In each of them two lines always meet, and this is contrary to Euclid's parallel axiom which states that through a point P not on a line L there is *only one parallel to L.*

Parallelism is a special property dealt with in an incidence geometry, called in the literature "affine" geometry ([96] Chap. 9).

In the affine system which follows, Axiom 2 of Axiomatic System II (Sec. 2.4) is replaced by a parallel axiom, and there are not five but six axioms. The effect of these changes is studied.

AXIOMATIC SYSTEM III (THE PAPPUS FINITE GEOMETRY)

Axioms
1. *If A_1 and A_2 are points of S, $\exists$ one and only one line containing both of them.*
2. *(The Parallel Axiom.) If P is a point not on a line L, $\exists$ exactly one line L′ through P parallel to L.*
3. *$\exists$ at least one line.*
4. *Every line contains exactly three points of S.*
5. *Not all points of S are on the same line.*
6. *If P is a point not on a line L, $\exists$ exactly one point P′ on L such that no line joins P and P′.*

A first consequence of Axiom 1 is:

Theorem 1 *Two lines of S meet in at most one point.*

and of Axiom 2 is:

Theorem 2 *Not all lines of S pass through a common point.*

and of Axioms 1, 4, and 6 is:

Theorem 3 *There are exactly three lines through each point of S.*

To determine the number of points in S, let points and lines be represented as in the previous section, and let the parallel through point 1 to line (4,5,6)

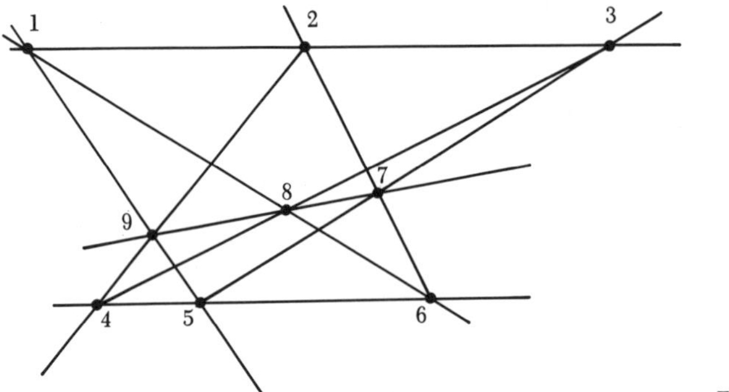

FIG. 2.3

contain the points 2 and 3 (Fig. 2.3). Also, let 2, 5 and 3, 6 be the pairs of points not joined by a line (Axiom 6), and let 7 be the third point of line (2,6). On the unique parallel through point 7 to line (4,5,6) are the two additional points 8 and 9. By Axiom 1 they are distinct from the points 1, . . . , 6; and S contains at least nine points and the nine lines represented by the number triples: (1,2,3), (4,5,6), (7,8,9), (1,5,9), (3,4,8), (2,6,7), (3,5,7), (1,6,8), (2,4,9).

Incidentally, there is a mechanical way to write these nine number triples. Consider the determinant

$$\Delta = \begin{vmatrix} 1 & 2 & 3 \\ 4 & 5 & 6 \\ 7 & 8 & 9 \end{vmatrix}$$

and its expansion

$$\Delta = 1 \cdot 5 \cdot 9 + 4 \cdot 8 \cdot 3 + 2 \cdot 6 \cdot 7 - 3 \cdot 5 \cdot 7 - 1 \cdot 6 \cdot 8 - 2 \cdot 4 \cdot 9$$

The six number triples appearing in this expansion, together with the three number triples

$$(1,2,3) \qquad (4,5,6) \qquad (7,8,9)$$

are the nine number triples representing the nine lines of Axiomatic System III.

Now to show that S contains exactly nine points, assume the contrary, and let a tenth point be 10. Then the point 10 is not on any of the lines just given, for each of them already contains the allotted three points (Axiom 4). Also, by Axiom 6, point 10 must be joined to each of two of the three points on a line. But if 10 is joined to a point, say 2, on line (1,2,3), then through 2 pass four lines contrary to Theorem 3, and from this contradition follows the next theorem:

Theorem 4 *There are exactly nine points in S.*

The somewhat trivial observation is made now that any two different axiomatic systems have in common all theorems whose proofs depend only on the axioms shared by the two systems. Thus Theorem 3 is shared by the Axiomatic Systems II and III of Secs. 2.4 and 2.5.

Axiomatic System III is noteworthy also for showing parallel lines, such as (1,2,3), (7,8,9), and (4,5,6), which have lost their familiar equidistance property.

Exercise 2.1

1 Does the figure consisting of the four points shown in Fig. 2.4 satisfy Axioms 1 to 3 of Axiomatic System I (Sec. 2.1)? Why?

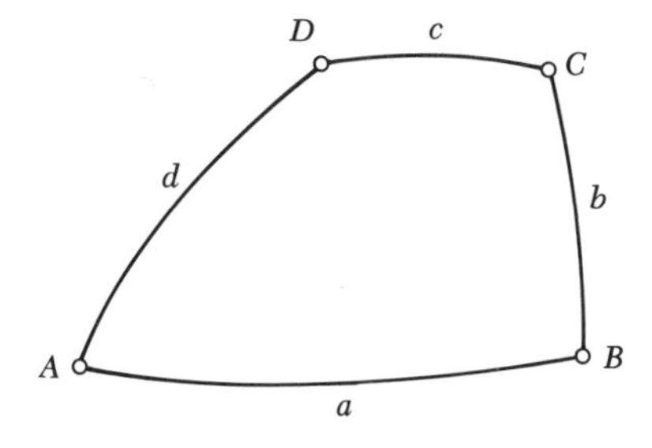

FIG. 2.4

2 Consider an axiomatic system based on the following axioms concerning points of a set S and subsets of them called lines:

> $\exists$ at least two points in S.
> If p and q are two points of S, $\exists$ one and only one line L of $S \ni p \in L$ and $q \in L$.
> If L is a line of S and p is a point of S not on L, $\exists$ one and only one line $L' \ni p \in L'$ and L' is parallel to L.

a. Does the system represent a finite geometry? Why?
b. Could books and library be taken as the respective models of points and line of this system? Why?
c. Deduce the two theorems: (1) "$\exists$ at least four distinct points in S." (2) "$\exists$ at least six lines in S."
d. Determine the exact number of points in S. *Hint:* See ([131] pp. 13–17).

3 Suppose that a dress manufacturer decides that on a certain new type of costume he is designing for spring trade, he will use combinations of the colors red, violet, blue, green, orange, yellow, and gray in the following way:

> The combination of any two colors should be used on one and only one costume.
> Any two costumes have one color in common.
> $\exists$ at least one costume.
> Every costume contains at least three colors.
> All the colors are not on the same costume.

Can one deduce from these statements the additional statement:

> *There are three colors which are not on the same costume.*

If not, give additional statements needed for the proof. *Hint:* Could there be one costume with six colors on it, the seventh color not being used?

4 a. Prove Theorem 2 of Axiomatic System II (Sec. 2.4).
b. Replace Axiom 4 of this system by the new axiom: "No line contains more than four points." How does this replacement affect the number of men in S? *Hint:* See ([45] vol. 1, p. 426).

2.6 The group axiomatically defined. Translations

In everyday life, the words "set" and "group" are often used interchangeably as when one speaks of a set or group of people, but in mathematics, the word is reserved for a very special meaning now to be explained. The group concept is introduced at this point to show how a set of axioms may be used to *define a new concept.*

Consider an axiomatic system whose undefined terms are elements a, b,

$c, \ldots$ of a set G. A binary operation denoted 0, on members of the set satisfies the four axioms:

1 For any $a, b \in S$, $a \, 0 \, b = c \in G$.
2 The operation 0 is associative: i.e., for all $a, b, c \in G$

$$(a \, 0 \, b) \, 0 \, c = a \, 0 \, (b \, 0 \, c)$$

3 For all $a \in G \; \exists$ an identity element $I \in G \ni a \, 0 \, I = a = I \, 0 \, a$.

4 Each element $a \in G$ has an inverse element $a^{-1} \in G \ni a \, 0 \, a^{-1} = I = a^{-1} \, 0 \, a$.

A group is now axiomatically defined: Any set of elements a, b, c, . . . satisfying these four axioms is said to form a group with respect to the operation 0.

The group is commutative if the additional axiom:

5 $a \, 0 \, b = b \, 0 \, a$

is satisfied. For example, the set J of positive and negative integers and zero form a commutative group with respect to the ordinary operation of addition. Zero is the identity element of the group, and the inverse of an element b is its negative, $-b$.

The positive integers *alone* do *not* form a group with respect to addition since there is no element x in this set such that $b + x = 0$ where b is any element of the set.

An important group in geometry is the set S of translations in the plane where by a translation is meant the process of moving a point $P(x,y)$ a given distance in a given direction to a point $P'(x',y')$.

If the lengths of the projections of the directed line segment PP' on the x and y axes are, respectively, h and k (Fig. 2.5), then

$$x' = x + h \qquad y' = y + k$$

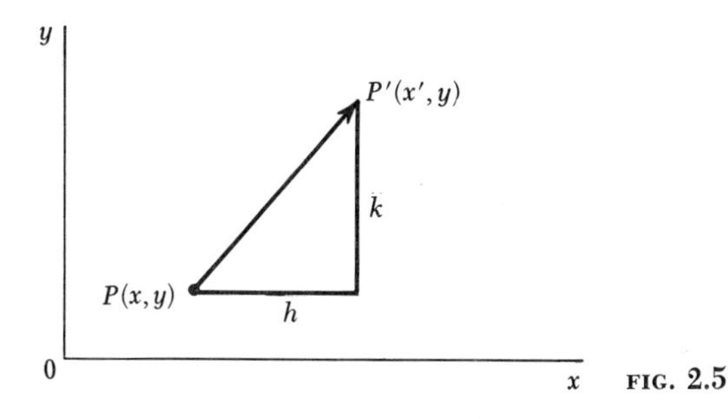

FIG. 2.5

By operation 0 on two elements T_1 and T_2 of S is meant one translation followed by another. Thus if T_1 and T_2 are the translations

$$T_1\colon x' = x + h_1 \qquad y' = y + k_1$$
$$T_2\colon x'' = x' + h_2 \qquad y'' = y' + k_2$$

$T_2 \; 0 \; T_1$ is the translation

$$x'' = x + (h_1 + h_2) \qquad y'' = y + (k_1 + k_2)$$

which belongs to the set.

The identity element of the set is the translation for which $h = k = 0$.

The associative axiom is satisfied since $T_1 \; 0 \; (T_2 \; 0 \; T_3)$ and $(T_1 \; 0 \; T_2) \; 0 \; T_3$ are both given by

$$x' = x + h_1 + h_2 + h_3 \qquad y' = y + k_1 + k_2 + k_3$$

The inverse of the translation $x' = x + h$, $y' = y + k$ is

$$x' = x - h \qquad y' = y - k$$

and thus all four axioms for a group are satisfied. Since also

$$T_1 \; 0 \; T_2 = T_2 \; 0 \; T_1$$

the group is commutative.

Exercise 2.2

1 Show that the set of all rational numbers, except zero, forms a group with respect to the operation of multiplication. Name the identity element of the set and the inverse of the element a.

2 Do the elements $1, -1, \sqrt{-1}, -\sqrt{-1}$ form a group with respect to the operation of multiplication? Give reasons for your answer.

3 Are the axioms for a group consistent? Why?

4 Consider the set of numbers 0, 1, 2, 3, 4 and an operation 0 on two of them, say a and b, where $a \; 0 \; b$ is (equals) the remainder when the ordinary product ab is divided by 5. For example, $304 = 2$ since $3 \cdot 4 = 12 = 2 \cdot 5 + 2$.
   a. Does the set form a group with respect to this operation? Explain.
   b. Name the inverse of each element of the set.

5 Consider the set G of six elements I, R, S, T, U, V[1]:

$$I = \lambda \qquad R = \frac{1}{\lambda} \qquad S = 1 - \lambda$$

$$T = \frac{1}{1 - \lambda} \qquad U = \frac{\lambda - 1}{\lambda} \qquad V = \frac{\lambda}{\lambda - 1}$$

[1] See in this connection Theorem 4.17.

each depending on a parameter λ. By the product of any two of these elements, say TU, is meant the function of λ obtained by substituting $U(\lambda)$ for λ in $T(\lambda)$. Thus

$$TU = \frac{1}{1-y} \qquad \text{where } y = \frac{\lambda - 1}{\lambda}$$

and hence

$$TU = \frac{1}{1 - \frac{\lambda - 1}{\lambda}} = \lambda = T$$

a. Verify the multiplication table below:

	I	R	S	T	U	V
I	I	R	S	T	U	V
R	R	I	T	S	V	U
S	S	U	I	V	R	T
T	T	V	R	U	I	S
U	U	S	V	I	T	R
V	V	T	U	R	S	I

b. Name the inverse of each element of G.

c. Show that the elements of G form a group.

d. Is the group commutative? Why?

6 Consider the rotations of an equilateral triangle ABC (Fig. 2.6) about its centroid M and its medians a, b, c, so as to transform the triangle into itself except for a possible permutation of the vertices A, B, C. If R_1, R_2, R_3 are the three rotations about M through the respective angles 120°, 240°, 360° and R_4, R_5, R_6 are rotations through 180° about the medians a, b, c, show that the set

$$\{R_1, R_2, R_3, R_4, R_5, R_6\}$$

forms a group with respect to the operation 0 where 0 is one rotation followed by another.

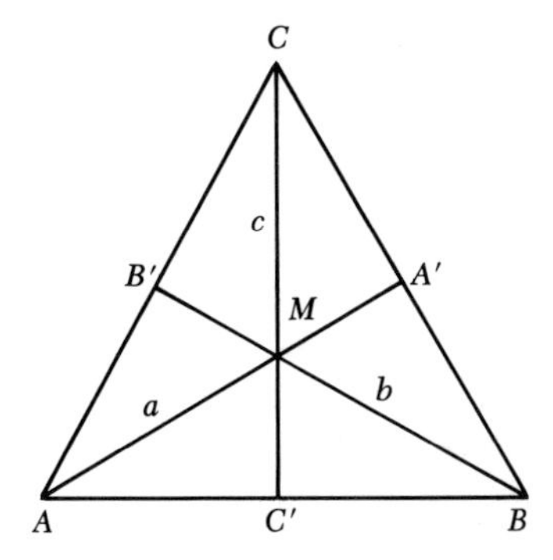

FIG. 2.6

2.7 Fields

The notion of a group is extended now to a mathematical system called a field. A set S of elements $a, b, c, \ldots$ is involved again, but this time there are *two* operations, addition and multiplication, denoted by the symbols $\oplus$ and $\odot$ to distinguish them from ordinary addition and multiplication.

The set S is said to form a field if (1) elements of S form a commutative group with respect to the operation: $\oplus$; and (2) when the identity element of this group is deleted the remaining elements form a commutative group with respect to the operation $\odot$; and (3) the distributive axiom is satisfied:

$$a \odot (b \oplus c) = [(a \odot b) \oplus (a \odot c)]$$

One of the most familiar examples of a field is the set of real numbers, with the two operations being ordinary addition and multiplication. Since

$$a + 0 = a \qquad \text{and} \qquad a \cdot 1 = a$$

0 is the identity element for addition and 1 for multiplication. Since also $a + (-a) = 0$ and $a \cdot 1/a = 1$, the inverse of a with respect to addition is $-a$ and with respect to multiplication is $1/a$ if $a \neq 0$.

Other examples of fields are the rational numbers a/b where a, b are integers and $b \neq 0$, and the complex numbers $a + bi$ where a, b are real numbers. The field of complex numbers contains the real numbers as a subfield since $a + bi$ is a real number when $b = 0$.

There are fields with only a finite number of elements. For example, the set of integers 0, 1, 2 forms a finite number field with respect to the two operations $\oplus$ and $\odot$ in which the results of ordinary addition and multiplication are replaced by their remainders when divided by 3. Thus, since $2 + 2 = 2 \cdot 2 = 4 = 3 + 1$,

$$2 \oplus 2 = 2 \odot 2 = 1$$

Complete addition and multiplication tables for the field are:

$\oplus$	0	1	2
0	0	1	2
1	1	2	0
2	2	0	1

ADDITION $\oplus$

$\odot$	0	1	2
0	0	0	0
1	0	1	2
2	0	2	1

MULTIPLICATION $\odot$

Exercise 2.3

1 Which of the axioms for a field are not satisfied by each of the following sets of numbers with the operations $+$ and $\cdot$ representing ordinary addition and multiplication?
   a. The even positive integers 2, 4, 6, . . .
   b. All fractions of the form $n/2$, n an integer
   c. The even integers, positive and negative, including zero
   d. All the integers

2 Which of the axioms for a field are not satisfied by the set of elements $\{0,1\}$ with the following addition and multiplication tables?

+	0	1
0	0	1
1	1	0

·	0	1
0	0	1
1	0	1

3 Consider the set of all real numbers with addition $\oplus$ and multiplication $\odot$ of two numbers a, b of the set defined as follows:

$$a \oplus b = a + b + 1$$
$$a \odot b = a + b + ab$$

Does the set form a field? Justify your answer.

2.8 Concerning the selection of axioms. Applied geometry

What dictates the axiom of an axiomatic system? In Axiomatic System I (Sec. 2.1), it was a physical ornament. In the system of Sec. 2.6 it was a desire to describe a new concept—the group. Euclidean geometry too had its practical origin. There now seems little doubt that Euclid's purpose in compiling his "Elements" was to derive properties of physical space from a few explicitely stated definitions and assumptions. To this end he started with 10 assumptions[1] concerning the undefined elements, point, line, and plane; but it was their physical counterparts, a dot, a ray of light, and a smooth mirrorlike surface, which helped him to formulate these axioms.

For example, the axiom which says that a straight line may be drawn between points is a highly idealized description of what happens when the point is replaced by a dot and a line by a taut string, ray of light, or rigid rod. While it is true that many physical lines may be drawn through two pencil dots, it is also true that if the dots become smaller and smaller, these different lines seem to coincide. By progressive abstractions, then, one finally

[1] See Appendix A, Sec. 2.3.

reaches a statement about elements which are stripped of all physical meaning.

An advantage of such a process should be obvious. Statements or axioms concerning undefined elements are no longer subject to those experimental errors made when dealing with physical objects, nor is the reasoning process distorted by what seems to be true of these physical elements.

It is not at all necessary that axioms of an abstract system describe properties of space or physical objects. Sheer fantasy, a powerful imagination, or a stroke of genius may be at work in the process of selecting axioms for a given system. If later, physical elements are found to have properties satisfying axioms of the abstract system, so much the better, for mathematics lives by virtue of its wide applicability. Theorems of the abstract system may then be used to describe additional and perhaps hidden properties of these physical elements. The abstract system is then said to be applied, and in such a case it is perfectly legitimate to speak of an axiom as a "self-evident fact" and of theorems as true, meaning that they are experimentally verifiable.

The application process has been illustrated in Boolean algebra, a highly abstract system which was constructed without reference to physical reality and later put to practical use in the construction of electric circuits and computing machines. The importance of these machines in modern research cannot be overestimated. The high-speed mathematical "brain," built at the Institute for Advanced Study by von Neumann, played a vital role in the hydrogen-bomb race. Calculations that would have required several lifetimes were made in a matter of months. In fact, it took this machine, believed to be the world's fastest and most accurate, six months to complete the computations on mathematical equations of the bomb.

2.9 Logical defects in Euclid. A paradox

Despite Euclid's unquestioned ability, his geometry does not satisfy the present-day requirements for logical rigor. It contains flaws. To begin with, tacit assumptions are not permitted in a logical system, and yet Euclid made quite a few of them. For example, he used the assumption that a line is infinite in extent in some of his proofs, but no such assumption was made in his axioms, nor is this property of a line a consequence of his other axioms. A line can be extended indefinitely in either direction, but this does not mean that the line is infinite in extent. A geometry will be studied later in which a line may be extended indefinitely in either direction and still be finite in length (see Sec. 3.13).

Again, in his proof by superposition, Euclid gives evidence of assuming facts not so stated in his axioms. The assumptions concerning congruence form an important part of any system of axioms for geometry. Lack of appreciation of this fact lies at the root of the difficulty involved in the method of superposition. In proving the congruency of two triangles having two sides and the included angle of the one equal to two sides and the included angle of the other, Euclid actually regarded one triangle as being moved in order to make it coincide with the other. He thereby tacitly assumed motion of figures without their deformation and completely ignored the fact that points are undefined elements. If, on the other hand, one considers geometry as an applied science in which figures are capable of displacement, or motion, there cannot be ignored the modern physical notion that the dimensions of bodies in motion are not the same as when they are at rest. Relativity theory has shown that space and time cannot be separated.

Furthermore, when Euclid constructed lines and circles to prove the existence of certain figures, he tacitly assumed their intersection points. In a rigorous development, the existence of these points must be either proved or guaranteed by means of an axiom. Euclid's failure to do this was later corrected by an axiom (see Appendix B, Sec. 6) which ascribes to all lines and circles that characteristic called *continuity.*

Another main defect in Euclid's system was his almost complete disregard of order concepts leading to such notions as the *two* sides of a line and the *interior of an angle.* Without the clarification of these ideas, absurd consequences result, such as the paradox:

Paradox *Every triangle is isosceles.*

The proof below makes use of theorems given in Appendix A. If a triangle ABC is isosceles, the perpendicular bisector of a side, say BC, and the bisector of the opposite angle A coincide, since in an isosceles triangle with $AB = AC$, the bisector of angle A is perpendicular to the base BC and bisects it. So if a triangle ABC (Fig. 2.7a) is not isosceles, the bisector of angle

FIG. 2.7

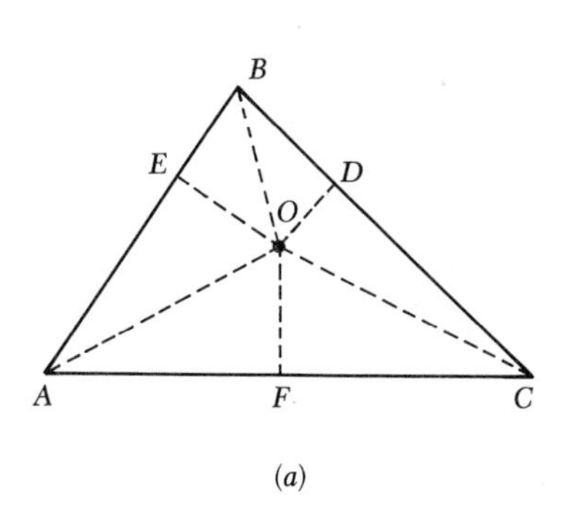

(a)

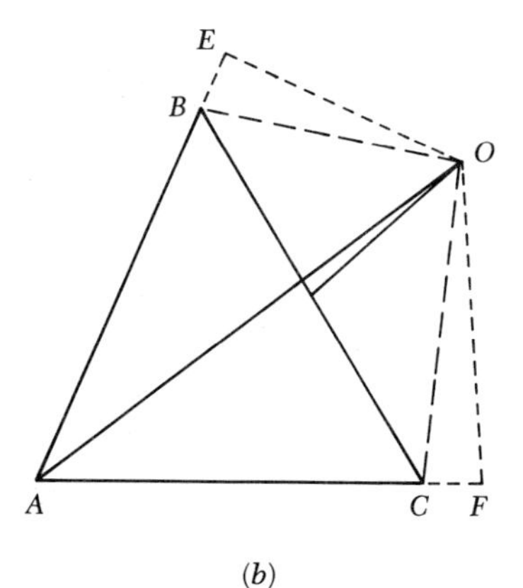

(b)

A and the perpendicular of bisector of BC are not parallel and hence meet at some point O within the triangle (Fig. 2.7a) or without the triangle (Fig. 2.7b).

In either case, the perpendiculars from O to the sides of angle A meet these sides in points E and F. Then the right triangles AOF and AOE have the hypotenuse AO in common and sides OE and OF are equal. The triangles are therefore congruent and

$$AE = AF \tag{1}$$

Also, since $OB = OC$ and $OE = OF$ the two right triangles EOB and FOC are congruent and

$$EB = FC \tag{2}$$

Hence, adding (1) and (2)

$$AE + EB = AF + FC \tag{3}$$

and subtracting (2) from (1)

$$AE - EB = AF - FC \tag{4}$$

But, in Fig. 2.7a, $AE + EB = AB$ and $AF + FC = AC$, and in Fig. 2.7b, $AE - EB = AB$ and $AF - FC = AC$. Hence, in both figures

$$AB = AC \tag{5}$$

and the $\triangle$s A are isosceles.

These arguments are valid consequences of Euclid's axiom but the situation calls for distinctions not possible on the basis of these axioms alone. Some order axioms needed to eliminate this paradox are given in Sec. 2.10. The reader is invited to study their role in this particular situation (see Exercise 2.4, question 1) and may consult the literature for further details ([39] p. 10). From these axioms it will follow that $AE + EB = AB$ *only when* E is *between* A *and* B and also $AE - EB = AB$ only when E is on AB extended.

Another proof with a flaw in it is given. This time it is the theorem of the exterior angle of a triangle (see Appendix A, Sec. 4, proposition 16).

To show that an exterior angle of a triangle ABC (Fig. 2.8) is greater than an interior angle, a point D is taken on BC extended, the midpoint of AC is called E, and line BE is extended to a point F such that $BE = EF$. Then $\triangle$s

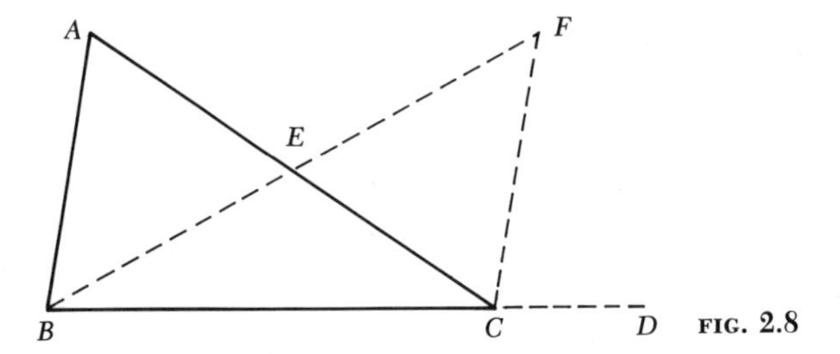

FIG. 2.8

BAE and ECF are congruent (Why?) and $\angle BAE = \angle ACF$. But $\angle DCA > \angle ACF$ and hence the exterior angle ACD is greater than the interior angle BAC. The flaws here are (1) the proof tacitly assumes that a line is infinite in extent and (2) a conclusion is based on the position of a point F in the figure. Geometries will be introduced later in which lines all have the same length and the position of F relative to other points of the figure is not as shown in the figure.

Exercise 2.4

1 In the nonisosceles triangle ABC (Fig. 2.7):
   a. Show that the bisector of $\angle A$ and the perpendicular bisectors of BC meet at a point O on the circumcircle of the triangle.
   b. If one of the two points D, E, say E, is between the two vertices A and B (Fig. 2.7a), the point D is not between B and C.

2 Name three theorems of Euclidean geometry based on the exterior angle theorem (see Appendix A, Sec. 4, proposition 16). Would these theorems hold if one rejects the assumption that a line is infinite in extent? Why?

3 Explain the fallacy in the following proof of the paradox: *A right angle is equal to an angle which is greater than a right angle.*

proof: Let $ABCD$ (Fig. 2.9) be a rectangle and E be a point outside the rectangle, chosen so that line segments AE and AB $(=CD)$ are equal. If perpendiculars at the midpoints H, K of segments CB, CE meet at point O, then in triangles OCD and OAE, $OD = OA$, $CD = AE$, and $OC = OE$. (Why?) These triangles are therefore congruent and $\angle CDO = \angle EAO$; and since $\angle ODA = \angle DAO$, $\angle CDA$ $(=90°) = \angle EAD$ $(>90°)$.

2.10 Another concept axiomatized. Order axioms

Another concept of elementary geometry is axiomatized. This time it is the much needed order concept, neglect of which led to paradoxical situations such as those discussed in the last section.

Pasch was among the first to characterize the relation among points on a

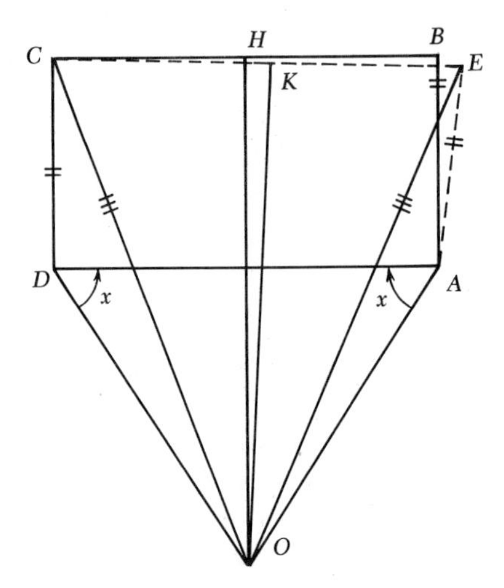

FIG. 2.9

line expressed by the familiar statement:" A point lies between two others of the line," but it is Hilbert's order axioms (Axioms 3.1, 3.2, 3.3, and 3.4, listed in Appendix B) that are being discussed here.

Axiom 3.1 states that *"If a point B is between A and C, then A, B, C are three different points of a line and B is also between C and A."* This axiom implies that the term "between" is used only for points on a line and states that the relative position of points A and C does not affect B's property of lying between A and C.

Axiom 3.2 states that *"Given points A and B on a line there exists at least one point C such that B lies between A and C."* This axiom guarantees the existence of at least three points on a line and allows one to refrain from setting stronger existence postulates in the incidence axioms (2.1, 2.2, and 2.3, Appendix B). (This last axiom guarantees the existence of three points, but not all on a line.)

The third order axiom (3.3) states that: "Given points A, B, C on a line, then at most one of these points lies between the two others." This axiom together with Axioms 3.1 and 3.2 permits the following definition:

> *An interval (line segment) consists of those points B of line AC for which B is between A and C, and B is called an interior point of the interval.*

The final order axiom (3.4) is Pasch's axiom: "If a line α intersects one side AB of a triangle (Fig. 2.10) in a point X between A and B and if α does not pass through C, then there exists on α a point Z between C and B, or a point Y between A and C."

From Axiom 3.1 it is seen that α cannot pass through A or B. But so far,

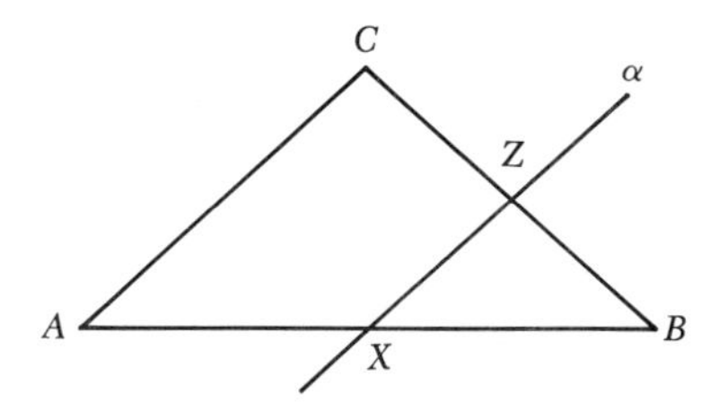

FIG. 2.10

it is not known whether there exists any point X between A and B. Axiom 3.2 assumes only the existence of a point on AB extended. The existence theorem below disposes of the question.

Theorem *If A and B are any two distinct points on a line, there exists a point X between A and B.*

proof: From Axiom 1.3, Appendix B, there exists a point E outside of the line AB (Fig. 2.11) and by Axiom 1.1, a line AE connecting the points A and E. Draw the line AE. According to Axiom 3.2, there exists a point F on AE and a point G on FB such that E is between F and A, and B is between G and F. Draw lines GE and FB.

Line GE does not pass through any of the points A, B, F. If it passed through A, point E would coincide with A, since EG intersects AF in E and therefore cannot intersect this line in a second point (Axiom 1.2); if GE passed through B, it would be line BGE and would intersect AF in F and not E. The same would be true if GE passed through F.

Now, since E is a point between A and F, Pasch's axiom may be applied to the triangle ABF. According to this axiom, GE must meet either line AB in a point X between A and B or line FB in a point Y between F and B. But the point Y cannot exist, since the intersection of lines FB and GE is, by construction, the point G; and by Axiom 3.3, point G cannot lie between F and B, since B lies between F and G. Therefore line GE intersects AB in a point X between A and B, and the theorem is proved.

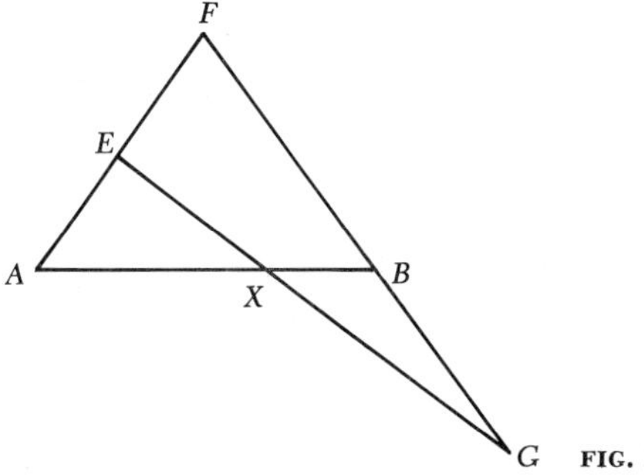

FIG. 2.11

Corollary 1 *If two points P and P′ are on the same side of a line α and if point Q is on the same side of α as P, then Q and P′ are also on the same side of α.*

proof: If line α (Fig. 2.12) were to intersect $P'Q$, it would follow from Pasch's axiom that it would also intersect PP' or PQ in an interior point.

Corollary 2 *If P and P′ are points on different sides of α and if point Q is on the same side of α as P, then P′ and Q are on different sides of α.*

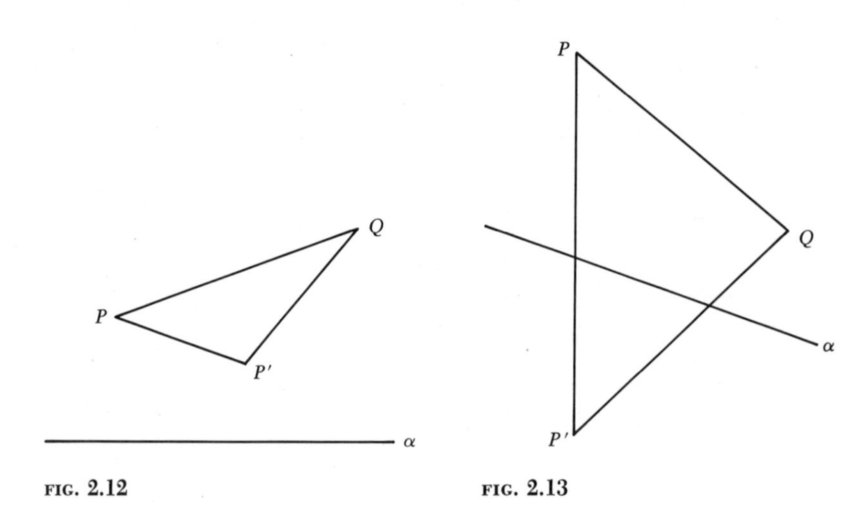

FIG. 2.12

FIG. 2.13

proof: By Pasch's axiom α must intersect QP' (Fig. 2.13) in an interior point, since, by hypothesis, it intersects PP' but not PQ.

Attention is called now to the distinction between Axiom 3.2 and the existence theorem just proved. In the former, a point C is assumed to exist on the line segment AB extended, and it can then be proved that there exists a point on AB between A and B. Originally the second order axiom read as follows:

If A and C are two points of a straight line, then there exists at least one point B between A and C and at least one point D so situated that C lies between A and D (Fig. 2.14).

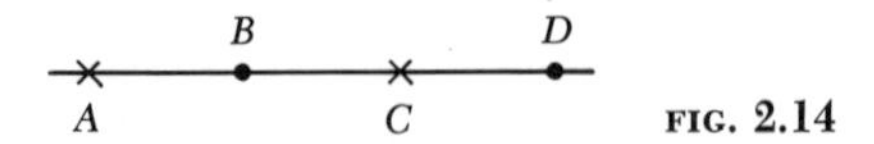

FIG. 2.14

In the light of the existence theorem just proved, this original axiom contained an assumption which was a logical consequence of the other axioms.

It was therefore removed, and the assumption became a theorem. This means that Hilbert's original set of order axioms was not independent.[1]

2.11 Comments on Hilbert's axioms and other sets for Euclidean geometry

Of the various sets of axioms which have been proposed for Euclidean geometry, Hilbert's, published in 1899, ranks high. For his undefined elements, Hilbert selected point, line, and plane and divided his axioms into the five subsets listed in Appendix B. The set is unusually large compared with certain others, such as Veblen's which contains only the two undefined terms "point" and "between" and 12 axioms.

Hilbert's assumptions in which congruence is taken as one of the fundamental undefined notions were probably suggested by the extensive controversies which took place as to whether Euclid regarded the idea of congruence or the idea of motion as fundamental. They are in marked contrast with those of the Italian Mario Pieri who took for his undefined terms *point* and *motion;* and with Birkhoff's set that uses *length* of a line and *angle measure* as primitive notions and with Prenowitz's which are built around *convex sets.*

One reason for the popularity of Hilbert's axioms stems from the fact that his set makes the derivation of fundamental theorems easier. It is, therefore, more closely related to the needs of elementary instruction than are others.

Hilbert's axioms are notable in another respect. They represent a conscious effort on Hilbert's part to choose a set of axioms with three special properties—*independence, completeness,* and *consistency*. These properties are defined and discussed in the next section.

2.12 Concerning consistency, completeness, and independence properties of a set of axioms

Axioms for a logical system are rarely set up just for the fun of it. But, even if one were tempted to do so, there is little probability that his system would be of any importance in either a mathematical or scientific sense unless the set was *consistent, complete,* and *independent.*

> *A set of axioms is said to be* consistent *if no two statements of the system contradict each other or if, of any two contradictory statements in the system, at least one cannot be proved.*
>
> *A set of axioms is called* complete *if, of any two contradictory statements involving terms of the system, at least one statement can be proved in the system.*
>
> *A set of axioms is said to be* independent *if it does not contain*

[1] See in this connection Sec. 2.12.

a single superfluous statement, i.e., a statement which can be deduced from the remaining axioms and which might therefore be counted among the theorems.

If the logical system is a geometry which is to be applied, it is important that axioms correspond well with properties of physical objects representing the undefined terms. In this way one arrives at theorems further describing these elements.

Of the three properties just defined, *independence* is the least important. If a set of axioms is not independent, there is simply redundancy in the system, and this may be desirable from both a psychological and pedagogical point of view. When, for example, a geometry is developed from a certain set of axioms, it may happen that an early theorem is difficult to prove. There would then be no objection to using this theorem as an axiom and pointing out later that it is not independent of others of the set as was done in Sec. 2.10.

As for the *completeness* property, its significance can be brought out by two simple examples. When Euclid selected his axioms, he obviously had in mind a set which would make it possible to deduce from them all the existing geometric facts as well as others that might be discovered in the future. In particular, the line was to contain an infinite number of points. A set of axioms such as Axioms 1 to 3 of Sec. 2.1 which resulted in a finite geometry would therefore be incomplete for a system in which the line contained an infinite number of points.

An even more significant example of an incomplete set of axioms is possible. When Euclid's parallel axiom is removed from his set, there remains what is commonly called "absolute" or "neutral" geometry ([96] Chap. 3). The deleted system is incomplete for a system which contains the theorem: "The sum of the angles of a triangle is 180° since proof of this theorem is based on the removed axiom."

Consistency is the one property which must be considered in more detail if the rapidly moving developments of the nineteenth and twentieth centuries are to be fully appreciated.

To link consistency with the completeness property, consider a system which contains both a statement p and its denial $\sim p$. Then the following sequence of statements is a valid argument:

$p \rightarrow (\sim p \rightarrow q)$	(tautology)	(1)
p	(hypothesis)	(2)
$\therefore\ \sim p \rightarrow q$	(modus ponens)	(3)
$\sim p$	(hypothesis)	(4)
$\therefore\ q$	(modus ponens)	

where p and q are any two statements. This means that from the hypotheses

p and $\sim p$ one may deduce any statement q. The resulting inconsistent system is therefore *complete*—but trivial. This is one reason for desiring the consistency of a set of axioms.

Consistency may also be linked with the independence property. To show that an axiom y of a system S is independent of the other axioms of the system, it is sufficient to show that these residual axioms and a new axiom $\sim y$ that directly contradicts axiom y form a consistent system S'. For, if axiom y were not independent of the other axioms of system S, it would be deduced as a theorem. The new system S' would then contain both axiom y and a contradiction of this axiom and hence be inconsistent.

Original confidence in the consistency of Euclid's early system was of course based on the belief that a model existed—and that model was physical space. But when it began to appear that non-Euclidean geometries (see Chap. 3) might also have physical space as a model, confidence in the model test for consistency was considerably shaken.

The Russell-Whitehead thesis that mathematics is only a chapter of logic seemed to advance the final solution of the consistency problem one further step until it was observed that this thesis simply changed the problem to a more general form. This then was the status of the consistency problem when in 1931, shortly after the publication of the "Principal," there appeared in the literature, a short paper entitled "On Formally Undecidable Propositions of Principia Mathematica and Related Systems." Its author Kurt Gödel, was at that time a young mathematician twenty-five years of age working at the University of Vienna. Since 1938 he has been a permanent member of the Institute for Advanced Study at Princeton, New Jersey, where he continues his investigations.

In his paper, Gödel presented the world with the astounding and quite disappointing conclusion that *it is impossible to establish the consistency of a large class of deductive systems unless one adopts principles of reasoning so complex that their consistency is as doubtful as that of the systems themselves.*

Gödel's paper is not for beginners but good summaries may be found in the literature [84].

Exercise 2.5

1 Using Hilbert's order axioms (Appendix B) prove the theorem: "Between any two points of a line lie an unlimited number of points."

2 Using Pasch's Axiom 3.4, Appendix B, prove the theorem: "If a line through a vertex A of a triangle ABC enters the triangle, it intersects the opposite side BC at a point between B and C."

3 In what sense are (a) the independence and consistency properties of a set of axioms related? (b) the completeness and consistency properties?

4 a. How may the independence of a set of axioms be established?

b. Show that each of the Axioms 1 to 5 of Sec. 2.4 is independent. *Hint:* See ([136] p. 47; [14] pp. 44–49).

5 What objection is there to showing the consistency of a set of axioms by means of a model. *Hint:* See ([84] Chaps. II and III).

6 An order relation, denoted $<$, is characterized by the set of axioms concerning elements a, b, c.

(o_1) If $a \neq b$, then $a < b$, or $b < a$.
(o_2) If $a < b$, then $a \neq b$.
(o_3) If $a < b$ and $b < c$, then $a < c$.

If the symbol $a < b$ means a is different from b:

a. Does axiom o_1 hold if $a < b$ is taken to mean "a is an ancestor of b"? Explain.

b. Is the set of axioms o_1, o_2, o_3 consistent? Explain. *Hint:* See [136].

7 Is the axiomatic system of Sec. 2.3 complete? Justify your answer.

Concluding remarks

Looking back over what has been done in this chapter, one sees the plan of starting with an analysis of simple axiomatic systems first and following them by others of increasing complexity. Axiomatic System I (Sec. 2.1) was a simple one with only two undefined terms and *three* axioms. Axiomatic System II (Sec. 2.4) had the same number of undefined terms, but this time there were *five*, not three axioms. Parallelism was introduced for the first time in the affine geometry of Axiomatic System III (Sec. 2.5), and after axiomatic approaches to order, group, and field concepts were given, comments were made on the axiomatic structure of Euclidean geometry.

The brief discussion of the three significant properties of an axiomatic system—independence, consistency, and completeness—is intended for use in Chap. 3 where non-Euclidean geometry is introduced.

The idea that there may be significant geometries other than Euclid's should now be just as natural and agreeable as the idea that two games, say football and basketball, can have two different sets of rules. What practical purpose each could serve is not so apparent.

For practical applications Euclidean geometry still holds a unique and enviable position in the hierarchies of geometries of the twentieth century. An actual flat triangular lot, for example, is assumed to possess the properties of its mathematical counterpart—a triangle in the Euclidean plane. So, in the triangular lot, as in the mathematical triangle, the greatest angle lies opposite the greatest side, the sum of its angles is 180°, and if the lot is a right triangle, the square of its hypotenuse is equal to the sum of the squares of its other two sides, as in a theorem of Euclidean geometry.

Are all such applications mute evidence of the Euclidean character of space? *The answer is no,* despite popular thinking to the contrary. If careful experiments should seem to indicate, for instance, that the hypotenuse of a right triangular lot is 5 yards when the other two sides are 3 and 4 yards, that would not be convincing evidence of the Euclidean character of space. Measurements are only approximations in which small errors cannot be detected. In fact, it is now known that Euclidean geometry is only a first approximation to the geometry of the physical universe. It is highly inadequate for many of the theoretical investigations of the twentieth century.

Suggestions for further reading

Blumenthal, Leonard M.: "A Modern View of Geometry."
Borzuk, K., and W. Szmielew: "Foundations of Geometry."
Coxeter, H. S. M.: "Introduction to Geometry."
Eddington, Sir Arthur S.: "The Nature of the Physical World."
______: "Space, Time and Gravitation."
Fetison, A. I.: "Proof in Geometry."
Heyting, A.: "Intuitionism: An Introduction."
Minkowski, H.: Time and Space, *Monist.*
Moise, E.: "Elementary Geometry from an Advanced Standpoint."
Nagel, E., and J. R. Newman: "Gödel's Proof."
Poincaré, H.: "The Foundations of Science."
______: "Science and Hypothesis."
Ramsey, F. P.: "The Foundations of Geometry."
Russell, Bertrand: "The Foundations of Geometry."
______: "Our Knowledge of the External World."
Wilder, R. L.: "Introduction to the Foundations of Mathematics."
Wylie, C. R.: "Foundations of Geometry."
Young, J. W.: "Lectures on Fundamental Concepts of Algebra and Geometry."

3 non-euclidean geometry

Any account of the discovery, rise, and development of non-Euclidean geometry reads like fiction. It is a completely absorbing tale of struggle, hardship, early defeat, and final success. When the story is told, the reader will see for himself how much the world is indebted to the men whose insight, skill, and vision brought unity into a vast collection of detached and apparently isolated theories. Quietly, steadily, and with determination, the early workers went about their colossal task of creating new geometries.

Early investigators were considerably hampered by Kant's philosophy of space. Even as late as the seventeenth century, Newton's laws, Euclidean geometry, and Kant's philosophy were still dominating the scientific world and progress was stymied.

Although he granted that knowledge of space could be obtained directly from experience, Kant felt that certain a priori knowledge existed in the mind, *independent of experience.* For instance, such a simple statement as "The straight line is the shortest distance between two points" was supposed to be a truth coming from "a priori judgment," and to Kant, this statement, like many of Euclid's axioms, was at once acceptable. Theorems were then but logical implications of the axioms. Yet, these theorems were at the same time supposed to describe the world of sense perceptions. Some of the arguments were certainly weak!

Fortunately, there were those who detected the weakness of such arguments, but for a long time these early critics were unable to gain any general support for their objections. As history shows, traditional habits of thought, stressed from infancy, are virtually impregnable walls, withstanding the assaults of advanced thinkers for long periods of siege. Eventually, however, the walls were weakened by pressure from all sides, and the world was forced to discard outmoded, erroneous beliefs. Some of the details of how this was done will show how much effort went into the project.

3.1 Historical background

If there is any one source to which the discovery of non-Euclidean geometry can be attributed, it is to the extensive studies and investigations of Euclid's axioms. Two of his axioms are of an essentially different nature from the others: (1) his parallel axiom, frequently called the fifth postulate, and (2) the axiom that a straight line is infinite in extent. Both involve an infinite concept and are, therefore, not experimentally verifiable. How, for instance, can one show by an experiment that a line is infinite in extent, when physical lines like stretched strings or rays of light always have finite length?

Because all the other axioms have a finite character and were suggested by their correspondence with physical objects, early geometers were led to investigate the question of whether or not the parallel axiom was dependent on the others. If so, it could be proved and, hence, would be a theorem, not an axiom, of the system.

Today, anyone who attempts to prove Euclid's parallel axiom is as much out of style as angle trisectors or circle squarers. In Euclid's day, however, and for hundreds of years thereafter, the brains and skill of the mathematical world were engaged in this problem.

Early investigators were, of course, hampered by Kant's philosophy and by the prevailing incorrect beliefs about the nature of space. That is why these investigators directed their efforts toward proving the dependence of the parallel axiom on the other axioms, rather than toward building geometries in which this axiom is denied.

One of the first attempts at a dependence proof was made by Proclus, a commentator on Euclid. He thought that he was dispensing with the parallel axiom when he defined the parallel to a given line as the locus of points at a given distance from the line. However, this merely shifted the difficulty, for it was then necessary to prove that the locus of such points was a straight line. Failing to do so, Proclus had to accept this fact as an axiom, and nothing was gained.

There were many others who attempted dependence proofs and failed, like Proclus, so that, by the beginning of the seventeenth century, the problem was still an unsolved and highly controversial one.

The first real progress in settling the question of the independence of the parallel axiom was made by Saccheri (1667–1733), an Italian Jesuit priest and a professor of mathematics at the University of Pavia. He conceived the brilliant idea of denying Euclid's parallel axiom, hoping thereby to deduce contradictory results and thus establish the truth of the axiom.

For his investigations Saccheri chose the birectangular isosceles quadrilateral $ABCD$ (Fig. 3.1) with angles A and B right angles and sides AD and BC equal. Then, by drawing diagonals AC and BD and using the congruence

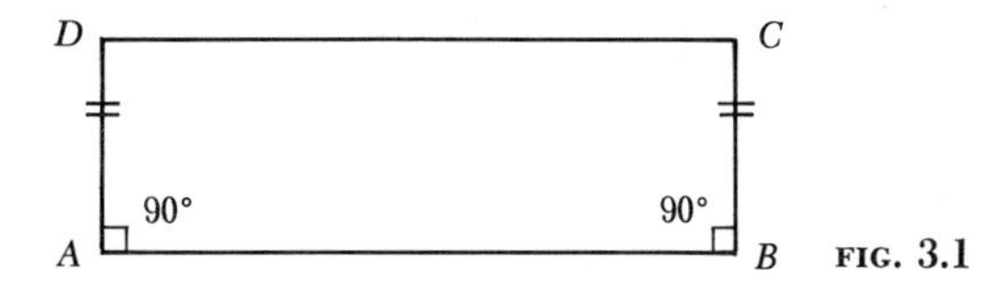

FIG. 3.1

theorems (Appendix A, Sec. 4), and not Euclid's parallel axiom, Saccheri established the equality of angles C and D. He then reasoned that the possibilities existed that the angles were acute ($<90°$), obtuse ($>90°$), or equal to 90°. Knowing that $C = D = 90°$ was a consequence of Euclid's parallel axiom, Saccheri assumed first that the angles at C and D were obtuse. From this assumption he deduced Euclid's parallel axiom, and since the latter implies $\angle A = \angle B = 90°$, the desired contradiction had been reached.

Saccheri next assumed that angles C and D were acute, hoping to arrive again at a contradiction. But it never appeared, despite the proof of one theorem after another. Had he realized that these theorems represented a perfectly logical, consistent geometry different from Euclid's, Saccheri would have anticipated the discovery of non-Euclidean geometry by at least a hundred years. Instead, his works went almost completely unnoticed until Beltrami, the distinguished Italian mathematician, convinced the scientific world of their worth.

The next investigations of note were those of the German mathematician J. H. Lambert (1728–1777). He took as his fundamental figure the quadrilateral with three right angles and examined the hypotheses in which the remaining angle was in turn right, obtuse, and acute. His results, like those of Saccheri, were indefinite and unsatisfactory. The investigations continued, however, and three outstanding periods followed.

It finally dawned on mathematicians of the late eighteenth and early nineteenth centuries that the unending record of failure in the search for proof of the parallel axiom was due not to any lack of ingenuity on the part of mathematicians but, rather, to the fact that the parallel postulate was independent of the other axioms. This meant that it was possible to replace the parallel axiom by a different one and, on the new axiomatic basis, to develop a logical system different from Euclid's.

The three mathematicians who finally broke with tradition and developed a geometry in which Euclid's parallel axiom was denied were Gauss (1777–1855) of Germany, Bolyai (1802–1860) of Hungary, and Lobachevski (1793–1856) of Russia. Bolyai's discovery, made in 1823, at the age of twenty-one, was not published until 1832; Lobachevski's was published in 1829; while Gauss, uncertain of his results and fearing the loss of prestige, failed to publish his results.

Although Gauss, Bolyai, and Lobachevski apparently worked independently of each other and in different parts of the world, each of them conceived the idea of replacing Euclid's parallel axiom by the new axiom that "through a point not on a line, there is more than one parallel to the line." The resulting geometry is the classic non-Euclidean system called hyperbolic geometry. It corresponds to the geometry originating from Saccheri's acute-angle hypothesis.

Unlike Bolyai, Lobachevski was a prolific writer, inspired, perhaps, by his great desire to gain world wide recognition for his geometry. Unfortunately, recognition failed to come during his lifetime, but as if to atone for its neglect, the world often refers to hyperbolic geometry as "Lobachevskian" geometry.

Once hyperbolic geometry had been developed and accredited, other non-Euclidean geometries appeared. Chief among these was elliptic geometry, first developed by Riemann (1826–1866) of Hanover, Germany, and described in his inaugural address, delivered in 1851, upon his admission as an unpaid instructor to the University of Göttingen.

In rejecting not only the parallel postulate but also the postulate that a straight line is infinite, Riemann broke with Euclidean geometry even more drastically than did his predecessors. He assumed that every line was endless and finite in length. His resulting geometry is that suggested by Saccheri's obtuse-angle hypothesis.

The possibility of a second kind of elliptic geometry was seen and developed later by Felix Klein (1849–1925) of Germany. His geometry and Riemann's are usually distinguished by designating Klein's as elliptic, and Riemann's as spherical, geometry. Some writers, however, prefer the respective names "single elliptic" and "double elliptic" geometry.

The second period in the development of non-Euclidean geometry is noteworthy for the many consistency proofs which appeared. The question of the consistency of a logical system had long been a troublesome, difficult, and challenging one. In the development of any logical system, there is always the burden of proving that the system is consistent by showing that no contradictions can occur. No contradiction has ever been found in either hyperbolic or elliptic geometry, but this fact does not constitute proof that these geometries are consistent.

Originally the consistency of a geometry was established by giving meaning to its undefined elements in terms of elements of Euclidean geometry. Then, axioms and theorems of the new system corresponded to theorems of this latter geometry, and the new geometry was seen to be as consistent as Euclid.

One of the first of these model tests for consistency was given by Beltrami in 1868 when he showed how hyperbolic geometry could be represented with suitable restrictions on a surface of constant Gaussian curvature. The test established not absolute but relative consistency.

There still remained the unanswered question: "Is Euclidean geometry consistent?"

To find the answer to this question, the mathematicians joined forces with the logicians. Inspired by the novel idea that mathematics is simply an extension of logic, Hilbert proposed a union of the axiomatic and logistic methods. By analyzing mathematical and logical processes and representing them in terms of an appropriate symbolism, such as the symbolic logic of

Chap. 2, Hilbert hoped to show that a formula and its contradiction could never be reached. Why his program could not be completely successful was shown later by Gödel.[1] The discovery of geometries as consistent as Euclid showed that Euclid's parallel axiom was not dependent on his other axioms; and thus investigations of this period brought to a close the great and magnificent efforts to prove the independence of Euclid's parallel axiom.

The third period in the development of non-Euclidean geometry is associated with the rise of a companion science—projective geometry—and its development without the aid of the distance concept. More will be said of this trend when projective geometry is studied.

Exercise 3.1

1 How many parallels are there to a line in (a) hyperbolic geometry, (b) elliptic geometry?

2 In which non-Euclidean geometry is a line boundless but finite in length? Give a model of such a line. Which of Hilbert's incidence axioms (Appendix B) are not satisfied by your model?

3 Given the birectangular, isosceles quadrilateral $ABCD$ (Fig. 3.1):
   a. Using Euclidean theorems not dependent on Euclid's parallel axiom, show that angles C and D are equal and that line EF joining the midpoint of the sides AB and CD is perpendicular to both lines.
   b. Using Euclid's parallel axiom and theorems of Euclidean geometry, prove that $\angle C = \angle D = 90°$.
   c. Using the hypothesis $\angle C = \angle D > 90°$, show that $CD \parallel AB$.
   d. In a geometry in which there is more than one parallel to a line, are angles C and D acute or obtuse? Why? If there is no parallel to a line, are these angles obtuse? Why?

HYPERBOLIC GEOMETRY

3.2 A new parallel axiom and some of its consequences

The following introduction to hyperbolic (Lobachevskian) geometry introduces this system as a modification of Euclid. It is based on Hilbert's axioms (Appendix B), together with the following replacement for Euclid's parallel axiom:

Axiom 5′ *Through a given point not on a given line, more than one line can be drawn not meeting the given line.*

The new axiom is not so unnatural as it might seem. Suppose that a ray through a point P, not on a line L (Fig. 3.2), rotates about P first in

[1]See Sec. 2.10.

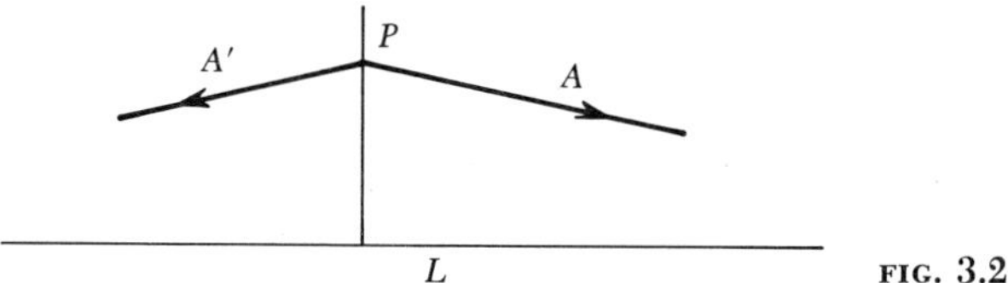

FIG. 3.2

the counterclockwise direction until it coincides with a nonintersecting ray PA and then in the clockwise direction until it coincides with another nonintersecting ray PA'.

Why should it be assumed, as is done in Euclidean geometry, that the rays PA and PA' coincide in the same line—the unique parallel PM to L? It is just as reasonable to assume that these lines are distinct.

Suppose that this latter assumption is made. What new theory results? A first consequence concerns the numbers of nonintersecting lines through P. Are there lines other than the lines PA and PA'? To answer the question, suppose L' and L'' are two distinct lines through P that do not meet L (Fig. 3.3). Then, if O is the foot of the perpendicular from P to L, one pair of vertical angles formed by the lines L' and L'' contains the line PO. Any line PX in the other (shaded) pair of vertical angles formed by these lines does not meet L, for if PX met L at a point M, then by Pasch's axiom (Axiom 2.4, Appendix B) the line L'', through the vertex P of the triangle POM, would meet L. But L'' is, by hypothesis, a nonintersecting line, and a contradiction has been reached.

Since PX is any line through P lying in the shaded vertical angles, it then follows that *there are an infinite number of lines through point P not meeting line L.*

Of this infinite set of nonintersecting lines, two particular ones will now be singled out for attention and called parallel lines.

As the line PO rotates about O in the counterclockwise direction, it intersects line L for a time and then ceases to intersect L. Consequently, lines through point P are divided into the two sets S_1, containing only intersecting lines, and S_2, containing only nonintersecting lines. From the continuity axi-

FIG. 3.3

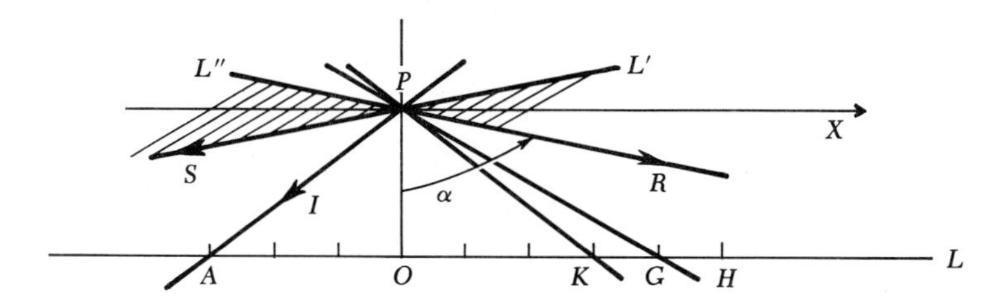

oms (Appendix B, Sec. 6) it then follows that *there is a line dividing one of these sets from the other.*

To which set does the dividing line belong? To answer this question, observe that the dividing line of the two sets must be either the last line of set S_1 or the first line of set S_2. But, there is no last line in set S_1, for if there were, it would meet L in a point G. Then, there would be a point H on L at a finite distance from O and on the opposite side of G from O. The line PH would then be an intersecting line, and PG would not be the last line of set S_1. This means that the dividing line of the two sets is a first line of set S_2. Call it PR. Similarly, another dividing line is encountered as PO rotates in the clockwise direction. Call it PS.

These two special nonintersecting lines, PS and PR, are called the respective *left and right handed parallels to L through point P.*

It will be shown next that the angles RPO and SPO which these parallels make with PO are equal.

Suppose that one of these angles, say SPO, is greater than the other. Then, there is a line PI forming with PO an angle α, equal to angle RPO. Let PI intersect line L at a point A. Then take a point K on the opposite side of O from A so that $OK = AO$ and draw OK.

Now, since the right triangles OPA and OPK are congruent, corresponding angles OPA and OPK are equal, and hence PK coincides with PR. But PK intersects L, and PR does not intersect L. From this contradiction follows the equality of the angles RPO and SPO. It is shown next that these angles are acute.

If $\angle RPO = \angle SPO = \alpha = 90°$, lines PR and PS coincide in the perpendicular through P to PO (Fig. 3.4), and there is then only one parallel to L through P, contrary to the hypothesis.

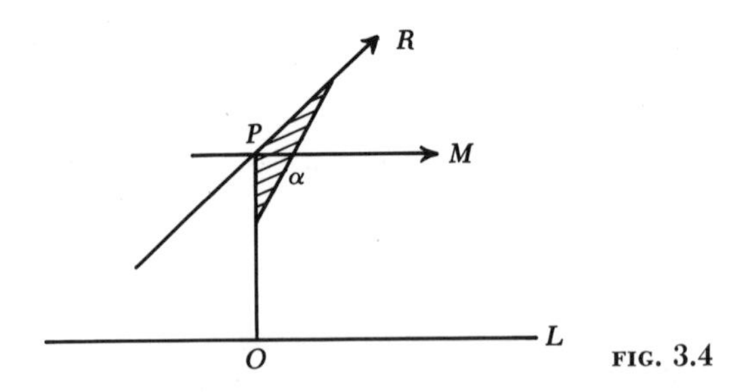

FIG. 3.4

If $\alpha > 90°$, then all lines lying in the shaded angle OPR (Fig. 3.4) are intersecting lines. But one of them is the perpendicular through point P to line L. It does not meet L (Why?), and again a contradiction has been reached. Thus proof has been given of the basic theorem:

Theorem 3.1 *If P is any point not on a line L, there are two distinct lines PR and PS which do not intersect L, and they form equal acute angles with the perpendicular PO to L. Any line through P in the domain bounded by PR and PS and not containing the perpendicular PO is a nonintersecting line. All other lines through P intersect L.*

Some of the strange consequences of the new parallel axiom will be exhibited by means of a model. In this way the somewhat tedious method of deducing theorems from axioms is avoided. Besides, a model at the start often suggests some of the queer results which could not have been anticipated by logical reasoning alone.

Fortunately, there are two famous models for hyperbolic geometry, one due to Klein, the other to Poincaré. Klein's model will be used primarily for illustrating the new theory of parallels and Poincaré's for illustrating some theorems directly contradicting Euclid.

Exercise 3.2

1 Show that the following are valid theorems in hyperbolic geometry:
   a. Two triangles are congruent if two sides and the included angle, two angles and the included side, or three sides of one triangle are equal to the corresponding parts of the other.
   b. In an isosceles triangle, the angles opposite the equal sides are equal.

2 Is it possible to prove without using the parallel axiom that an exterior angle of a triangle is greater than either of the interior and opposite angles? Give reasons for your answer.

3 Prove the hyperbolic theorem: "The sum of two angles of a triangle is less than 180°."

3.3 Klein's model

Klein's model for the real hyperbolic plane is constructed by taking some familiar elements of elementary geometry and renaming them in such a way that all but one of Euclid's axioms are satisfied. The one exception is the famous parallel axiom.

In the model, the (real) hyperbolic plane consists of points interior to a given circle C (Fig. 3.5). These interior points are distinguished from ordinary points of the plane by being called H-points. Chords of the circle are then taken as models of lines and are distinguished from ordinary lines by being called H-lines or hyperbolic lines. All points outside the circle are ignored.

Since two points within the circle determine a chord and since two chords intersect in one and only one point, two H-points determine an H-line and

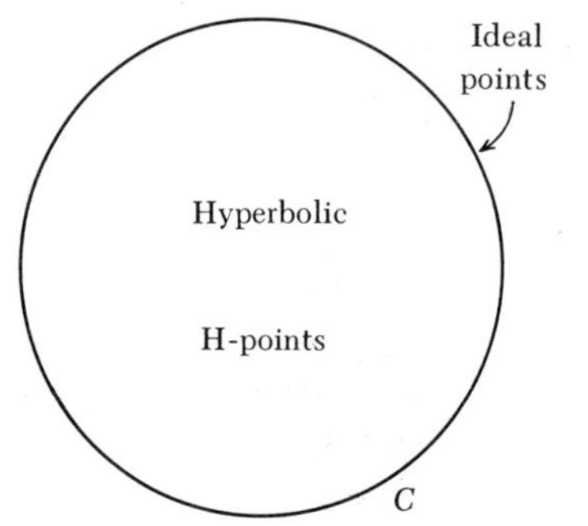

FIG. 3.5

two H-lines intersect in an H-point. Thus Hilbert's incidence axioms (Appendix B, Sec. 2) are satisfied by elements of the model.

Also, if chord AB meets circle C in points O and U (Fig. 3.6) and X is an H-point, the lines XU and XO do not meet line AB, for (by assumption) points on the circumference do not belong to the set of H-points. The lines XO and XU are, therefore, models of the respective left and right handed parallels to line AB. Actually seeing a model of two parallels to a line forces the reader to discard some old-fashioned ideas about how parallels should look. The H-lines XU and XO of Fig. 3.6 are parallels, not because of their appearance in a drawing, but because of the new definition of parallel lines.

Klein's model is satisfactory for illustrating the new theory of parallels but not for visualizing or anticipating many of the new theorems. However, it has served a far more useful purpose. Throughout the Middle Ages and up to the beginning of the nineteenth century, all efforts to prove the dependence of Euclid's parallel axiom on his other axioms had failed. Klein's model showed the inherent impossibility of such a proof. If the parallel axiom could be deduced from Euclid's other axioms, it would be a theorem in the geometry of Klein's model, and proof has just been given that it is not.

Mathematical genius of a rare order went into the finding of the model. Once it was found, hyperbolic geometry was seen to be as consistent

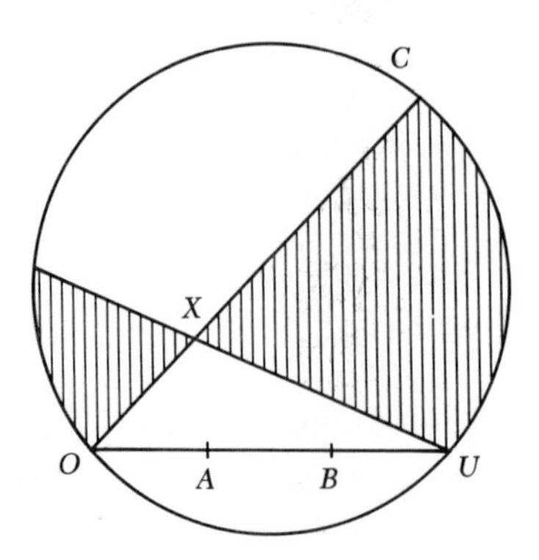

FIG. 3.6

as Euclid's since if both a theorem and its contradiction could be deduced in the new science, the same would be true in Euclidean geometry.

Exercise 3.3

1 Why are the first 28 theorems of Euclid, Book I, valid theorems of hyperbolic geometry?

2 Does a (hyperbolic) line through point X of Fig. 3.6, lying within the shaded region, intersect (hyperbolic) line AB? Give reasons for your answer.

3 Prove that if one line of the hyperbolic plane is parallel to a second, the second is parallel to the first. *Hint:* See ([133] p. 69).

3.4 A geometry of orthogonal arcs. Inversion theory

In Poincaré's model of hyperbolic geometry (Sec. 3.7) circular arcs orthogonal[1] to a fixed circle are models of hyperbolic lines. To show why this is so, it will be necessary to digress temporarily to study the Euclidean geometry of these orthogonal arcs. Some of their properties will be established and used to illustrate strange theorems of hyperbolic geometry. Inversion theory appears in the study.

Definition *Two points P and P' of the plane are said to be inverse with respect to a given circle, center O, radius r, if $OP \cdot OP' = r^2$ and also if both points are on the same side of O.*

The symbol $(O)_r$ is used for the circle with center O and radius r, or simply the symbol (O) when the radius r is immaterial. Circle $(O)_r$ is called the circle of inversion and the transformation which sends point P into P' is known as an *inversion*.

According as $OP \gtreqless r$, $OP' \lesseqgtr r$, and thus the effect of this particular transformation is merely an interchange of the inside and outside of the circle with points of the circle remaining fixed. Also, the one point which has no inverse is the center O of the circle of inversion, for when P is at point O, $OP = 0$ and the relation $OP' = r^2/OP$ is meaningless.

The *inverse point P' to P is easily constructed.* When point P is inside the given circle $(O)_r$ (Fig. 3.7), a perpendicular to OP at P meets it at points T and T', and tangents at these points to the circle intersect at the inverse point P' to P. From the similarity of triangles OTP and OTP' follows the relation

$$\frac{OP}{OT} = \frac{OT}{OP'}$$

and hence $OP \cdot OP' = \overline{OT}^2 = r^2$.

[1] Two intersecting circles are orthogonal if their tangents at a point of intersection are perpendicular.

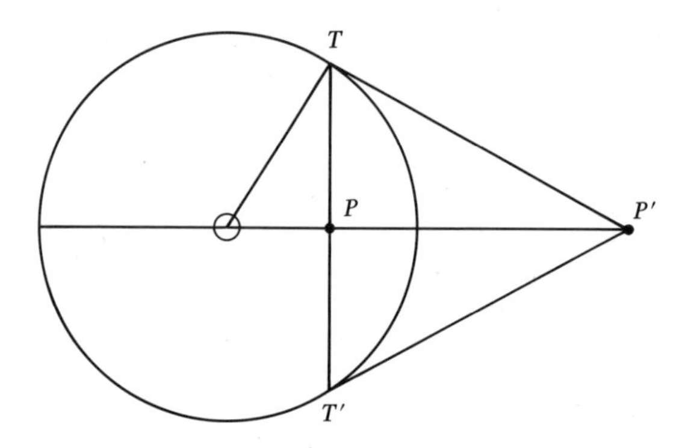

FIG. 3.7

The *construction when P is outside the circle* simply reverses the process: If tangents from P to circle $(O)_r$ meet the circle at T and T', the inverse point P to P' is the intersection of lines TT' and OP.

A particularly significant fact concerning inverse points and orthogonal circles is stated formally in the next theorem:

Theorem *Any circle through a pair of points, inverse with respect to a given circle, is orthogonal to this circle, and conversely any circle cutting the given circle orthogonally and passing through P also passes through its inverse point P′.*

proof: Let (C) be the circle through the points P, P' inverse with respect to circle $(O)_r$ (Fig. 3.8), and let line OC meet this circle in points Q and Q'.

Then, if the two circles intersect at point M,

$$OQ \cdot OQ' = OP \cdot OP' = OM^2 = r^2$$

or, since $OQ = OC - CQ$ and $OQ' = OC + CQ$,

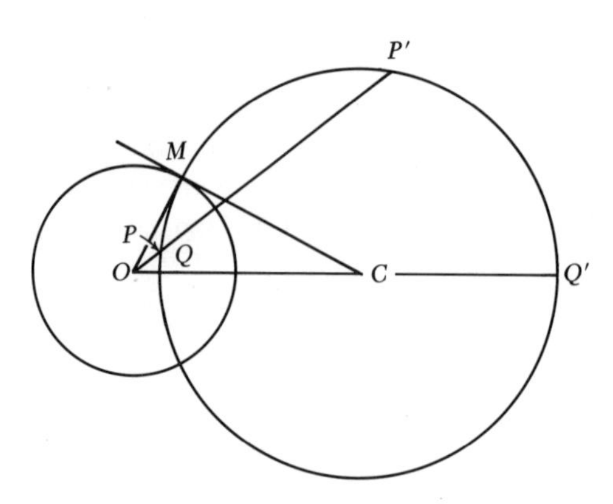

FIG. 3.8

$$\overline{OC}^2 - \overline{QC}^2 = \overline{OM}^2 \quad \text{or} \quad \overline{OM}^2 + \overline{MC}^2 = \overline{OC}^2$$

from which it follows that $\angle OMC = 90°$. The circles (O) and (C) are therefore orthogonal.

Conversely, when $\angle OMC = 90°$ and OP meets the circle (C), through P, in a second point X,

$$OP \cdot OX = OM^2 = r^2 = OP \cdot OP'$$

and X is the inverse point P' to P.

Corollary 1 *Through two interior points A, B of a circle (C), there is one and only one circle orthogonal to circle (C).*

This is so because the circle through A, B passes also through the inverse points A', B' of A and B with respect to (C). (See Exercise 3.4, question 1.)

Corollary 2 *Two orthogonal circles which intersect in two distinct points and which are both orthogonal to a third circle have one point of intersection inside, and the other outside, the third circle.*

This is so because if the circles intersect at a point P, their second point of intersection is the inverse point P' to P; if a point is inside a circle (O), its inverse with respect to (O) is outside (O).

Exercise 3.4

1 Show that if A, A' and B, B' are two pairs of inverse points, a circle through three of the points passes through the fourth point.

2 If two intersecting circles are both orthogonal to a third circle, show that their points of intersection are collinear with the center of the third circle.

3.5 Inverse curves

As point P moves on a curve C, its inverse point P' moves on a curve C' called the inverse of C.

The four theorems below list the inverses of lines and circles according as they do or do not pass through the center of the circle of inversion.

Theorems *1* *A line through O inverts into a line through O.*
2 *A line not through O inverts into a circle through O.*
3 *A circle through O inverts into a line not through O.*
4 *A circle not through O inverts into a circle not through O.*

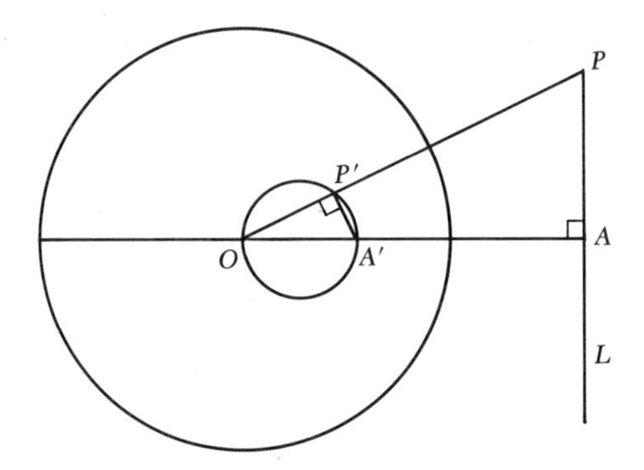

FIG. 3.9

Theorem 1 is an immediate consequence of the fact that the inverse point P' to P lies on the line OP.

To prove Theorem 2, let A be the foot of the perpendicular from O to the given line L (Fig. 3.9), and let A' be its inverse point. Then

$$OA \cdot OA' = r^2$$

and since P' and P are inverse points,

$$OP \cdot OP' = r^2$$

The triangles OAP and $OA'P'$, therefore, have a common angle and including sides proportional, and hence they are similar. Therefore

$$\angle OAP = \angle OA'P' = 90°$$

But the arc in which a 90° angle is inscribed is a semicircle. Point P' therefore lies on a circle whose diameter is OA', and Theorem 2 is proved.

A reversal of these arguments gives the proof of Theorem 3.

To prove Theorem 4, let point A (Fig. 3.10) be the center of the given circle, and let a line through O intersect OA in points P and Q. If P' and Q' are respective inverses of P and Q,

$$OP \cdot OP' = r^2 = OQ \cdot OQ' \tag{1}$$

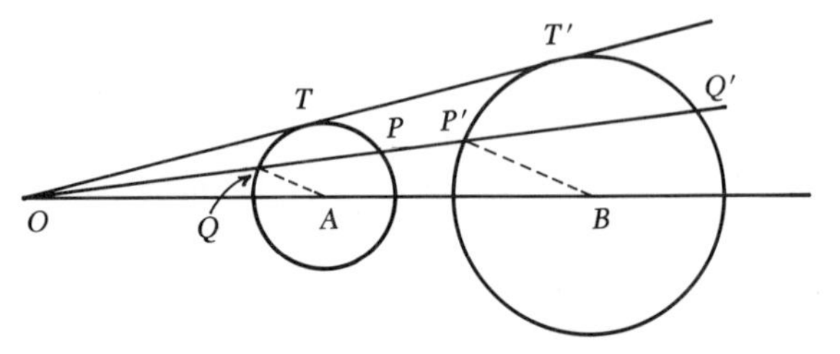

FIG. 3.10

Now, from Theorem 61 of Appendix A the tangent OT from O to circle A satisfies the relation

$$OP \cdot OQ = (OT)^2 \tag{2}$$

and from (1) and (2) it then follows that

$$\frac{OP'}{OQ} = \frac{OQ'}{OP} = \left(\frac{r}{OT}\right)^2 \tag{3}$$

Next, draw a line through P' parallel to QA, meeting OA at point B. Then triangles OQA and $OP'B$ have three angles of one equal to three angles of the other and hence are similar. Therefore

$$\frac{OP'}{OQ} = \frac{OB}{OA} = \left(\frac{r}{OT}\right)^2 = \frac{P'B}{QA} = \text{a constant, say } K$$

This means, since QA is a constant, $P'B$ is a constant. Also, since OA is a constant, $OB =$ a constant, and B is a fixed point on line OA. Point P' therefore lies on a circle with center B and radius $= K \cdot \overline{QA}$, and Theorem 4 is proved.

3.6 Invariants

That angles do not change on inversion is established next.

Theorem *The angle between any two curves intersecting at a point different from the center O of inversion is unchanged under inversion.*

proof: Let the given curves C and D (Fig. 3.11) intersect in a point P distinct from the center O of inversion, and let any line through O

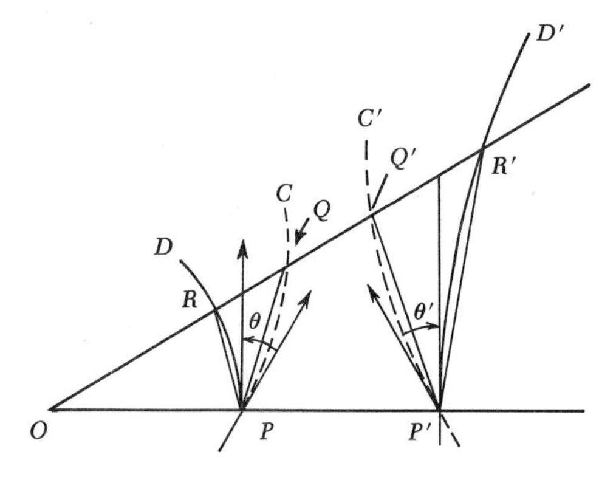

FIG. 3.11

other than OP intersect these curves in the respective points R and Q. The inverse curves to C and D, namely, C' and D', then intersect at the inverse point P' to P. Furthermore, if curves C' and D' are met by line OR in the points Q' and R', inverse to the respective points Q and R, the theorem will be proved by showing that the angle θ between the tangents at P to curves C and D equals the angle θ' between the tangents at P' to curves C' and D'.

Since triangles OPQ and $OP'Q'$ are similar, as are also triangles RPO and $R'P'O$,

$$\angle QPO = \angle OQ'P' \tag{1}$$

and

$$\angle RPO = \angle OR'P' \tag{2}$$

Hence, subtracting (2) from (1),

$$\angle QPR = \angle Q'P'R'$$

Now, as the line OQ rotates about O, the secant lines RP and PQ approach as limits the respective tangents at P to curves C and D. Similarly, the secant lines $Q'P'$ and $R'P'$ have as their limits the respective tangents as P' to curves C' and D'. Since angles QPR and $Q'P'R'$ are equal for every position of the rotating line, the limit angles θ and θ' are equal, and the theorem is proved.

Although inversion preserves the magnitudes of angles, it reverses their sense. In other words, as the ray PQ generates the angle QPR in a counterclockwise direction, its inverse ray $P'Q'$ generates the angle $Q'P'R'$ in the clockwise direction.

Since orthogonal circles intersect at a 90° angle and tangent circles at a 0° angle, three corollaries of this theorem are:

Corollary 1 *Orthogonal circles invert into orthogonal circles.*

Corollary 2 *Tangent circles not through the center of inversion invert into tangent circles.*

Corollary 3 *Circles tangent to each other at the center of inversion invert into a system of parallel lines.*

Exercise 3.5

1 If a circle is inverted into a circle, will the center of the first be the inverse of the center of the second? Why?

2 If A, B are two points within a circle $(O)_r$, construct the circular arc A, B orthogonal to $(O)_r$ and construct also the inverse of a line through O.

3 Construct the inverse of circles (A), (B), (C), each tangent to line L at the center O of inversion.

4 If two circles are orthogonal, show that the inverse of the center of either circle with respect to the other circle is the midpoint of their common chord.

5 Let arcs AC, BC, AB, each orthogonal to a fixed circle $(O)_r$, be inverted in a circle whose center R is the second point of intersection of arcs AC and BC (Fig. 3.12). Show that arcs AC and BC invert into straight lines $A'C'$ and $B'C'$, while arc AB inverts into an arc $A'B'$ of a circle.[1] Construct each inverted curve.

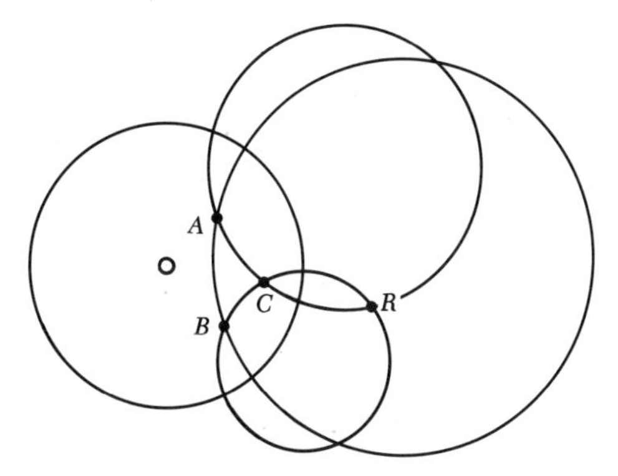

FIG. 3.12

3.7 Stereographic projection and Poincaré's model

Because Poincaré's model for hyperbolic geometry will be obtained by a stereographic projection, some elementary facts about such a projection are given first.

Suppose a sphere tangent at M to a horizontal plane p is projected from its highest point N on the plane (Fig. 3.13). Then, to each point P' of the sphere corresponds the point P in which line NP' meets plane p. The single exception occurs when point P' is N. This map of the sphere $P \to P'$ is called a stereographic projection, and two properties of such a projection stated without proof are:

Property 1 *Circles on the sphere not through N are mapped into circles of the plane, and conversely every circle in the plane is the image of a circle on the sphere.*

Property 2 *Stereographic projection reproduces angles on the sphere without distortion.*[2]

[1] See in this connection Figs. 3.15 and 3.16.
[2] For proofs see ([59] pp. 248–254).

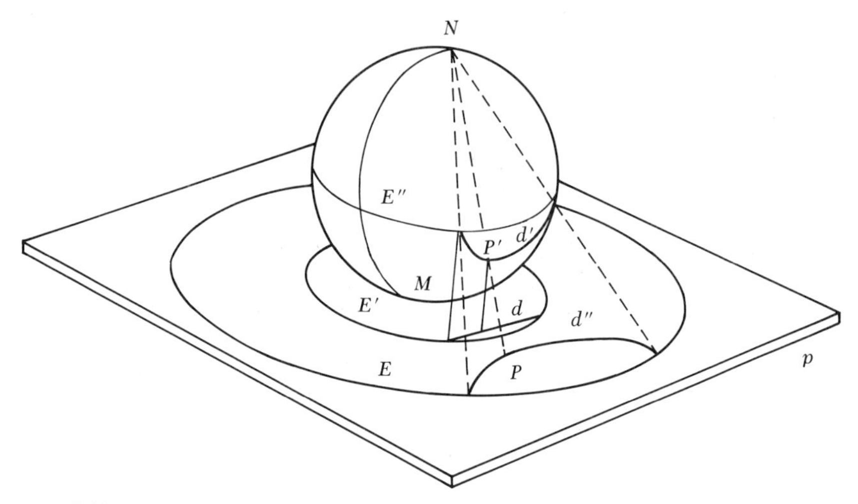

FIG. 3.13

Poincaré's model will now be described. The hyperbolic plane is first represented by the interior of circle E' lying in the horizontal plane p of Fig. 3.13. A sphere of the same radius r as E' is tangent to this plane at the center M of E'. The circumference and interior of E' are projected by a vertical parallel projection on the lower half of the sphere and its boundary circle E''. Then, each chord d of circle E' projects into a semicircle d' orthogonal to circle E''. If interior points of E' are points of Klein's model, the semicircle d' is the image of an H-line of the latter.

Now, map the lower hemisphere, whose boundary is E'', back on the horizontal plane by a stereographic projection from the point N (Fig. 3.13). By property 1, circle E'' maps into the circle E of the plane p, and each point of the hemisphere is mapped into an interior point of this circle. Because of the angle-preserving and circle-preserving nature of stereographic projection, semicircles d' of the hemisphere are mapped into circular arcs d'' orthogonal to circle E, and thus to each chord (H-line) of circle E' corresponds an arc orthogonal to circle E. Instead of studying the geometry of Klein's model, one may therefore study the geometry of points and orthogonal arcs of the fixed circle E. They are Poincaré's model of points and lines of the hyperbolic plane. Note, in particular:

1 Two interior points of circle E determine one and only one arc orthogonal to circle E.
2 Two orthogonal arcs intersect in one and only one point of circle E.

These statements will be recognized as incidence axioms of hyperbolic geometry if the word "line" is substituted for the words "orthogonal arc."

Here then is a situation in which a model of a line is not straight but curved.

3.8 Distance and angle in Poincaré's model

The distance AB between two points A and B of circle E of Poincaré's model is given by the definition

Definition *If the orthogonal arc joining points A and B of circle E meets this circle in points O and U, the distance AB between the points is given by the formula:*

$$AB = k \log \frac{OA/AU}{OB/BU}$$

where OA, AU, OB, BU are lengths of chords of the orthogonal arc AB and the real number $k \neq 0$ is a parameter, the choice of which determines the unit of length.

The formula is fantastic and yet it does have the familiar property that "the distance between two points is zero when the points coincide," for when $A = B$, the formula gives $AA = k \log 1 = 0$. The formula also reveals an *unfamiliar* property of a line. By the usual limit processes, the distance AB approaches $-\infty$ as B approaches U, and similarly the distance AO approaches ∞ as B approaches O. Unlike the Euclidean line, the hyperbolic line has therefore *two distinct* infinitely distant points.

Since orthogonal arcs XU and XO (Fig. 3.14) meet orthogonal arc AB at infinitely distant points, they are models of the respective right and left handed parallels to the line.

Another beauty of Poincaré's model is its conformal nature. The angle be-

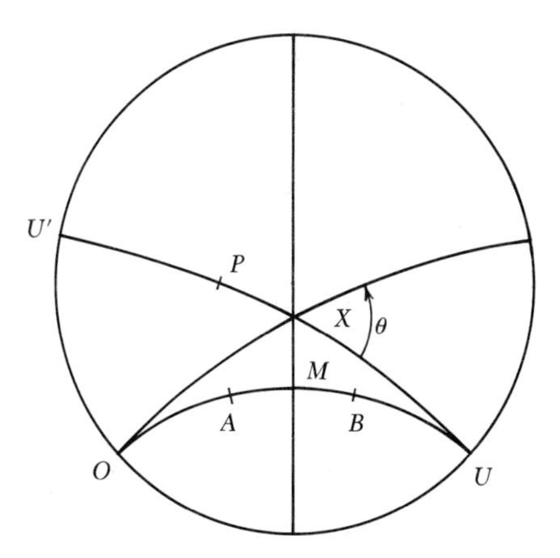

FIG. 3.14

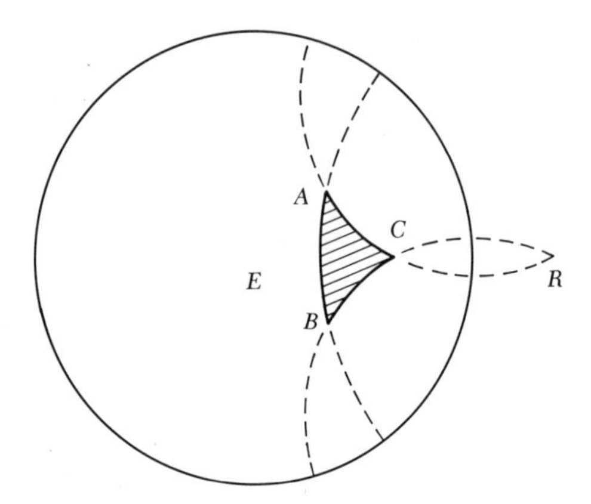

FIG. 3.15

tween two hyperbolic lines is by definition the angle θ between the two orthogonal arcs which model them. Thus the figure formed by arcs AC, CB, BA, each orthogonal to a fixed circle (Fig. 3.15), is a model of a hyperbolic triangle whose angles are given by corresponding angles between tangents at a vertex to arcs through the vertex. This fact is used in establishing a hyperbolic theorem directly contradicting Euclid.

3.9 Concerning the sum of the angles of a triangle

In the triangle ABC of Fig. 3.15, each of the circular arcs AC, BC, and AB is orthogonal to the fixed circle (E), and the arcs AC, BC meet at a second point R. Invert the entire figure in the circle having R as a center.

By Theorem 3 of Sec. 3.5, arcs AC and BC invert into the intersecting lines $A'C'$ and $B'C'$ (Fig. 3.16); and by Theorem 4 of the same section, the fixed circle (E) inverts into a circle orthogonal to lines $A'C'$ and $B'C'$ and hence into a circle with center at C'.

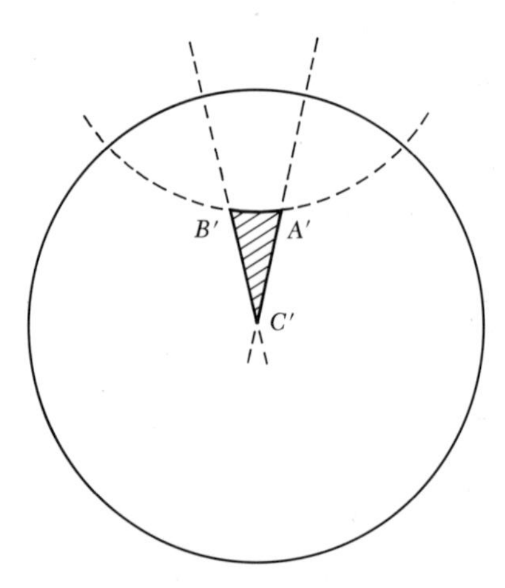

FIG. 3.16

Since the circular arc AB inverts into the circular arc $A'B'$ orthogonal to circle (C'), its center is outside this circle. But the triangle $A'B'C'$ of Fig. 3.16, formed by two straight lines and this orthogonal arc, has an angle sum less than that of the triangle whose vertices are A', B', C' and whose sides are all straight lines. Since the latter triangle has its angle sum equal to 180°, the triangular figure $A'B'C'$ bounded by two straight lines and a curved line has an angle sum less than 180°. But, by the theorem of Sec. 3.6, angles are unchanged on inversion, and hence the same is true of the original triangle ABC. Thus proof has been given of the next hyperbolic theorem:

Theorem 3.2 *The sum of the angles of a triangle is less than 180°.*

This means that a theorem of Euclid has been contradicted in hyperbolic geometry.

Since a quadrilateral consists of two triangles, each of whose angle sum equals the sum of the angles of the quadrilateral, another hyperbolic theorem is:

Theorem 3.3 *The sum of the angles of a quadrilateral is less than 360°.*

3.10 The Saccheri and Lambert quadrilaterals in hyperbolic geometry

Distance is used in describing the Saccheri quadrilateral $ABCD$ of Fig. 3.17. This figure is a quadrilateral whose angles at A and B are right angles and whose sides BC and AD are of *equal length. Angles C and D are then called the summit angles of the quadrilateral,* and a theorem concerning such a quadrilateral is:

Theorem 3.4 *The summit angles of a Saccheri quadrilateral are equal and acute.*

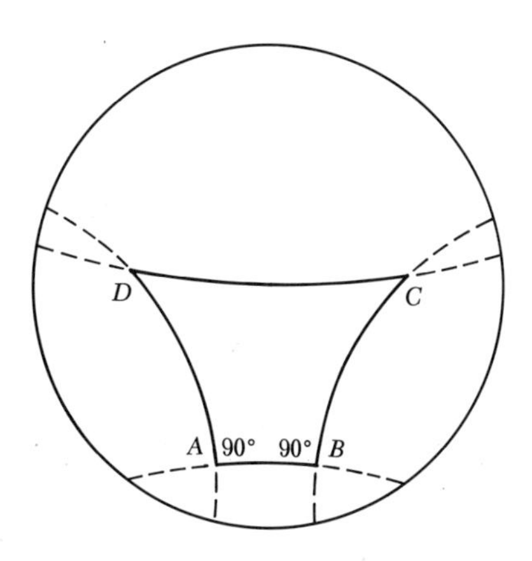

FIG. 3.17

The theorem is proved easily on the basis of the congruency theorems and Theorem 3.3 ([133] pp. 77–79).

The historic Lambert quadrilateral also has a queer property. Since three of its angles A, B, C are right angles (Fig. 3.18), its fourth angle D is acute by virtue of Theorem 3.3. Consequently, an almost incredible hyperbolic theorem is:

Theorem 3.5 *There are no squares in hyperbolic geometry.*

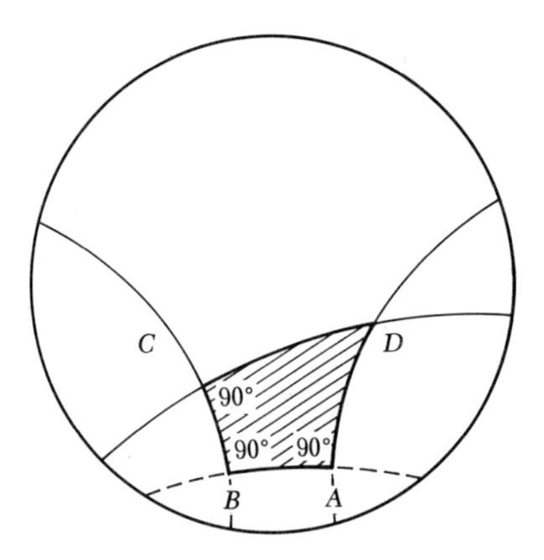

FIG. 3.18

Similar triangles have also lost a familiar property in the new science. If two similar triangles, such as ABC and $A'B'C'$ (Fig. 3.19), have their corresponding angles A and A', B and B', C and C' equal, and if these two triangles are not congruent, then the quadrilateral $BB'C'C$ has an angle sum of 360°, contrary to Theorem 3.3. Hence another unfamiliar theorem is:

Theorem 3.6 *Two similar triangles are congruent.*

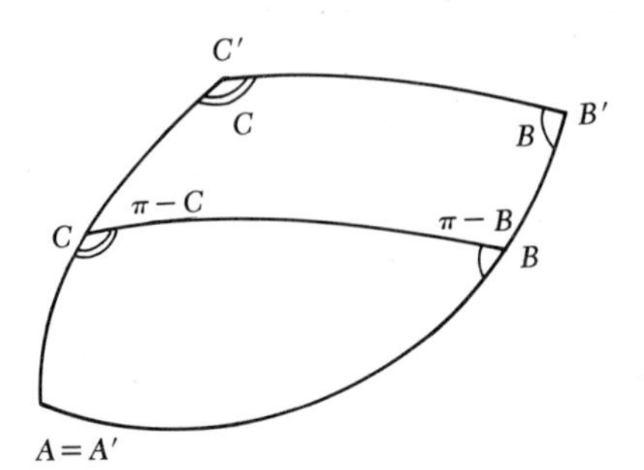

FIG. 3.19

3.11 The pseudosphere. Other facts

Many other strange facts are illustrated by figures on a pseudosphere (Fig. 3.20), the surface formed by revolving a curve called the tractrix about an asymptote. The tractrix may be described roughly as the path in the plane of a heavy point attached to one end of a taut string whose other end moves along a straight line in the plane. In Fig. 3.20 the tractrix has OX as an asymptote, and the tangent to the curve at the point P meets OX in a point A, where the length of segment PA is a constant.

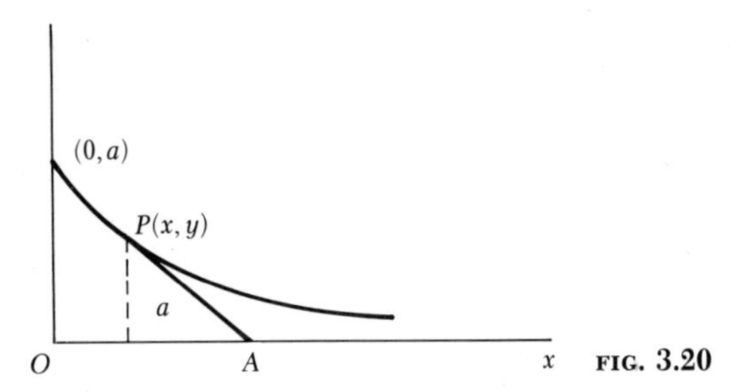

FIG. 3.20

Positions BC and AD (Fig. 3.21), taken by the tractrix while it is being rotated, are meridians of the pseudosphere. They are models of parallel lines. Even though the lines are converging toward each other, they will never meet. Also, on this surface the Saccheri quadrilateral $ADCB$ with its two equal acute summit angles no longer seems like the figment of a wild imagination.

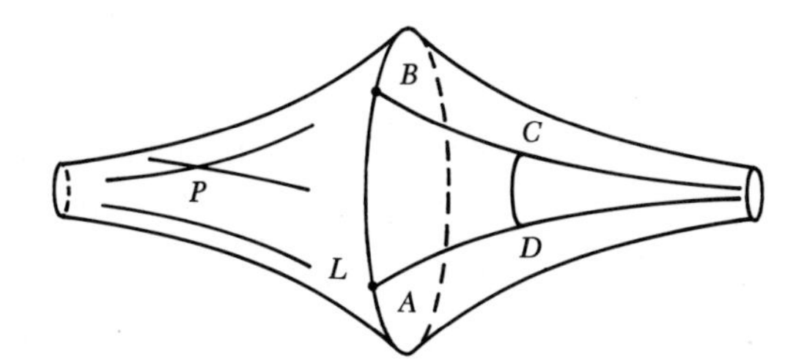

FIG. 3.21

Unfortunately, a study of this model takes one out of the realm of elementary mathematics and hence must be reserved for more advanced investigations.

Exercise 3.6

1 Prove that in a Lambert quadrilateral the sides adjacent to the acute angle are greater than their respective opposite sides.

2 Which is greater, the base or the summit of a Saccheri quadrilateral? Prove your answer.

3 Prove that the segment joining the midpoints of two sides of a triangle is less than one-half the third side.

4 Prove that two Saccheri quadrilaterals with equal summits and equal summit angles, or with equal bases and equal summit angles, are congruent.

3.12 Hyperbolic geometry and the physical universe

Originally, hyperbolic geometry represented a very excellent exercise in logic, but once man had freed himself of outmoded beliefs, his imagination soared to the point where he actually conceived the idea of applying the new science to the physical universe.

Lobachevski, observing that Euclidean trigonometric formulas are valid in the infinitesimally small neighborhood of a point of the hyperbolic plane, considered the possibility of his geometry replacing Euclid's in astronomical space. A crucial experiment, he thought, would consist in finding a positive lower bound for the parallax of stars. His experiment was doomed to failure, since such a lower bound, if it exists, is smaller than the allowance for experimental error. Failure of the experiment would merely mean that, if space is hyperbolic, the absolute unit must be many million times as large as the diameter of the earth's orbit.

It is not difficult, however, to imagine a universe subject to hyperbolic laws. Suppose, for example, that, in a huge sphere of astronomically large radius r, the temperature T changes from point to point in accordance with the law

$$T = c(r^2 - h^2)$$

where c is a constant and h is the distance of a point P from the center O of the sphere. The temperature T then has a maximum value for h equal to zero, is equal to zero for h equal to r, and is constant on any concentric sphere.

Under the assumption that people and objects in this fanciful world are so susceptible to the heat that they grow large or small with the temperature, it is possible to show that the shortest distance path joining two points within the sphere is along an arc of a circle orthogonal to the sphere. Thus, if a plane through the center of the sphere intersects the sphere in the circle C, the shortest distance curve joining two points A, B within C is along the arc orthogonal to C. Also, if this arc meets circle C in the points O and U (Fig. 3.14), arc AB is the shortest distance path for a person walking to the boundary point U of his world. As this person walks toward U, the distance from the center of the sphere increases, and hence his body shrinks and his steps become shorter. He would therefore get the distinct impression that he could never reach the boundary, no matter how many steps he would take. Nor would measuring instruments help him, for all objects are assumed to contract as their distance from the center increases. The observer's world would therefore seem to be infinite in extent.

If now the earth is placed in this sphere, so that an observer at point X is very near the center of the sphere, the shortest distance paths through X are circular arcs XO and XU orthogonal to circle C. Furthermore, if the radius of the sphere is sufficiently large, these arcs would look like straight lines, and the angle θ between these arcs would then be so small that no instrument would be fine enough to detect a value for it other than zero. The parallels XO and XU would then seem to coincide in a single line, and the observer would doubtless conclude that through the point X there is one and only one parallel to the line AB, as in Euclidean geometry. The inadequacy of visual observation or of measurements for determining the character of space is obvious.

ELLIPTIC GEOMETRY

An investigation is to be made next of elliptic non-Euclidean geometry in which Euclid's parallel axiom is replaced by the new axiom: "There are no parallels to a line."

Credit for the initial development of this system goes to the inimitable German mathematician Bernhard Riemann (1826–1866), who in 1854 read a dissertation to the Philosophical Faculty at Göttingen describing his new creation. Completely ignoring the parallel assumptions of both Euclidean and hyperbolic geometry, Riemann began his system with what was *at that time* the unbelievably fantastic assumption: "Lines are boundless but finite in length." He knew that some new concept of a line was needed because parallels will always exist when a line is infinite in extent.

Since great circles on a sphere are closed curves, all of the same length, the sphere immediately presents itself as a model surface for elliptic geometry if its points and great circles are identified with points and lines of the abstract system.

Because great circles always intersect, two lines always meet, and so there are no parallels to a line. However, the fact that great circles always intersect in *two* distinct points, not one, is a disturbing factor. Also, since there are infinitely many great circles through two diametrically opposite points of the sphere, two points do not always determine one and only one line. This means, even though the parallel axiom is satisfied, that incidence axioms are not satisfied, and the transition from Euclidean to elliptic geometry cannot be accomplished as easily as that from Euclidean to hyperbolic geometry.

3.13 The separation concept. Single and double elliptic geometry

The assumption that a line is infinite in extent is usually used in a proof of the Euclidean theorem: "Two lines perpendicular to the same line are parallel."

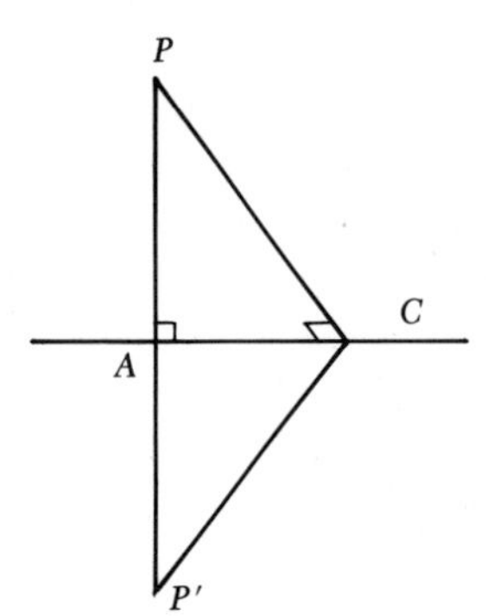

FIG. 3.22

There is another proof in which this assumption is not used. Suppose that the distinct perpendiculars to a line L at points A and C (Fig. 3.22) are not parallel and hence meet in a point P. If PA is extended its own length to a point P', the two triangles APC and $AP'C$ are congruent and hence

$$\angle ACP' = \angle ACP = 90°$$

But then the straight line PCP' having two points P and P' in common with the perpendicular PA coincides with PA, and the perpendiculars at A and C are not distinct, contrary to hypothesis.

Unlike the proof given in elementary texts, this one does not make use of the infinite extent of a line but rather of the separation principle: "A line L of the plane divides points not on L into the two sets S_1 and S_2 such that $S_1 \cap S_2 = \phi$."

On the assumption that every line of the plane separates the plane into two parts having no common point, the points P and P' of Fig. 3.22 *are distinct. Otherwise, the proof fails.*

Two courses of action are therefore possible: (1) abandon the separation principle and assume that two lines intersect in only one point (as in Euclidean geometry); (2) retain the separation principle, but assume that two lines, such as NAS and NBS (Fig. 3.23), intersect in not one but *two* points.

The first course of action leads to *single* elliptic geometry and the second to *double* elliptic geometry. In either system N, or S, is called a pole of the line AB. The names *double* and *single* elliptic geometry are suggested by the fact that a line has two poles in double elliptic geometry and only one pole in single elliptic geometry.

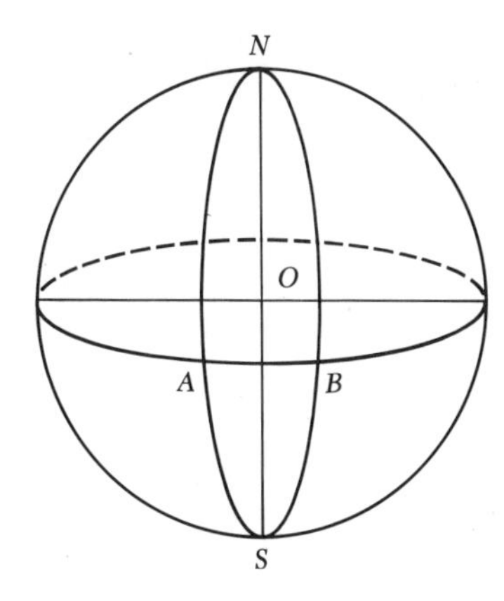

FIG. 3.23

3.14 The Möbius leaf

Another fact is observed. The elliptic plane in double elliptic geometry has the character of a two-sided surface such as a sphere on which the two sides are usually distinguished by calling one the inside and the other the outside. (The sphere does nct have an edge.) In Klein's single elliptic geometry, on the other hand, the plane has the character of a one-sided surface, a simple example of which is the Möbius leaf. This surface which has a most unusual separation property is described next.

Take a long, narrow rectangular strip of paper $ABCD$ (Fig. 3.24), and form a surface by joining the opposite sides AB and CD after first rotating one of them, say CD, about its midpoint through an angle of 180°. Corners C and D will then coincide (Fig. 3.25) with corners A and B, respec-

FIG. 3.24

FIG. 3.25

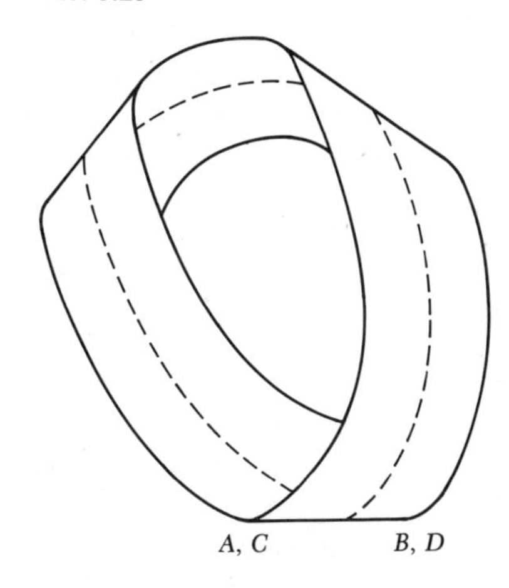

tively, and two sides of the surface will become indistinguishable. In other words, the surface has one side.

One way of distinguishing a one-sided surface from a two-sided one is to fix a point on the surface and a direction of rotation about it. If a point describes a closed path upon a two-sided surface, when the point returns to its initial position, the final direction of rotation coincides with the initial one. On a one-sided surface, however, there exist closed paths for which the original and final directions of rotation are opposite. This can be seen by drawing a line PQ in the middle of the rectangle $ABCD$ and noting the path on the resulting Möbius surface. A bug crawling along the PQ will return to its original position upside down. To see why, take a pencil with an eraser at its top and let the pencil describe the path PQ. If at the start the pencil is in the position NP (Fig. 3.26), then at the end, the eraser will be at point N' of the same figure such that $\angle NPN' = 180°$. Thus N and N' are on opposite sides of P.

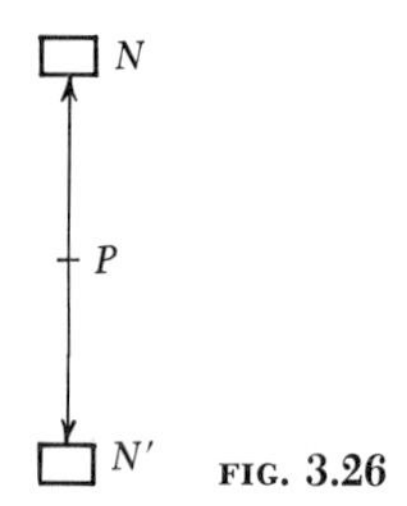

FIG. 3.26

A painter who contracts to paint a Möbius strip will find, when he has returned to his starting line, that he has done twice as much work as he had intended to do. Your model will explain why.

Exercise 3.7

1 If N (Fig. 3.23) is the pole of the great-circle arc AB of a sphere and the arc NA is extended its own length to a point S, what theorems of Euclidean geometry are needed to prove that the points S, B, and N are on a line? Do these theorems depend on either Euclid's parallel axiom or the infinite extent of a line? Explain.

2 Name a theorem of Euclidean geometry which does not depend on either the parallel axiom or the infinite extent of the line. Is your theorem a theorem also of elliptic geometry? Why?

3 If A and B are two points of a sphere, how many great circles are there through these two points if neither point is an extremity of a diameter of the sphere? Why?

4 a. Cut the Möbius leaf along the line PQ (Fig. 3.24). Into how many parts does the cut divide the surface?

   b. If the cut is made along the line LM (Fig. 3.24), where rectangle $ALMD$ is one-third of the rectangle $ABCD$, interpret the separation results on the Möbius leaf.

5 How many edges has the Möbius leaf? Give reasons for your answer.

3.15 Models

Now that a distinction has been made between single and double elliptic geometry, models will be given for each.

A model for single elliptic geometry is obtained by taking only half the sphere, a hemisphere with its boundary circle C, and identifying its points and great-circle arcs with points and lines of the plane.

Since elliptic lines are closed, each point A of circle C must be identified with its diametrically opposite point A' so that a great-circle arc with extremities at A and A' (Fig. 3.27) represents a closed line.

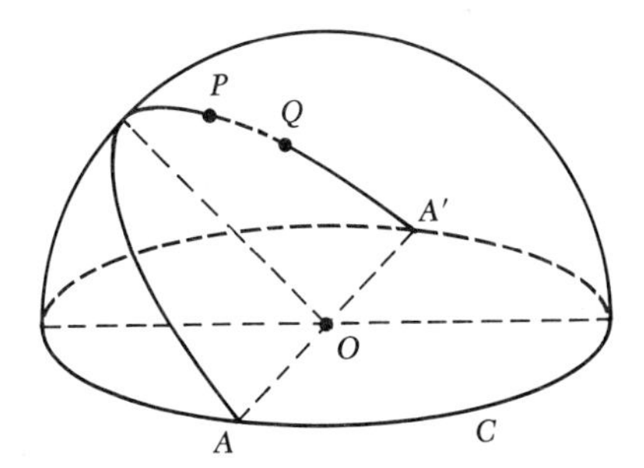

FIG. 3.27

The metric of the model is that of the sphere itself. Also, if PQ denotes any segment of a line, it is not known which of the two segments determined by P and Q is meant. Thus, in Fig. 3.27, segment PQ could be the dotted segment or the heavy segment containing the point A, now identified with its diametrically opposite point A'.

The angle θ between two lines of the elliptic plane is taken to be the angle between corresponding arcs representing these lines, and the angle is always understood to be less than 180°. This agreement is necessary in order that two sides and an included angle will determine one and only one triangle on the surface.

A model for double elliptic geometry may be taken to be the sphere itself. In contrast to the situation in single elliptic geometry, the line in double elliptic geometry has two poles, and instead of the axiom that "two lines intersect in a point," there is the following theorem:

Theorem 3.7 *Two lines intersect in two points and enclose a region of finite area* $A \neq 0$.

This theorem is illustrated in Fig. 3.23, where the great-circle arcs NAS and NBS are boundaries of a portion of the surface of the sphere.

A number of theorems common to single and double elliptic geometry are stated and illustrated next.

3.16 Basic theorems

The great-circle arcs NA and NB of Fig. 3.23 are models of perpendicular lines in the elliptic plane and illustrate the following theorem:

Theorem 3.8 *Two perpendiculars to a line meet in a point.*

The proof of the theorem follows immediately from the axiom that two lines of the plane always meet. (There are no parallels.)

For visualizing some of the following theorems, it will be convenient to think of the earth's surface as a perfect sphere. Its meridians are then models of lines perpendicular to the equator and since these meridians all meet at the north and south poles, there is illustrated the next theorem.

Theorem 3.9 *The perpendiculars at all points of a line are concurrent in a point called the pole of the line, and conversely every line through the pole is perpendicular to the line.*

The distance q from a pole of a line to the line is, by definition, the polar distance to the line, and a theorem concerning polar distance is:

Theorem 3.10 *In any triangle ABC in which angle C equals $90°$, angle A is less than, equal to, or greater than $90°$ according as the segment BC is less than, equal to, or greater than the polar distance q.*

The theorem is illustrated in Fig. 3.28, where $\angle B = \angle C = 90°$, angle $A < 90°$, and $BC < q$. Then $\angle A + \angle B + \angle C > 180°$, and the same figure also illustrates the next theorem:

Theorem 3.11 *The sum of the angles of a triangle is greater than $180°$.*

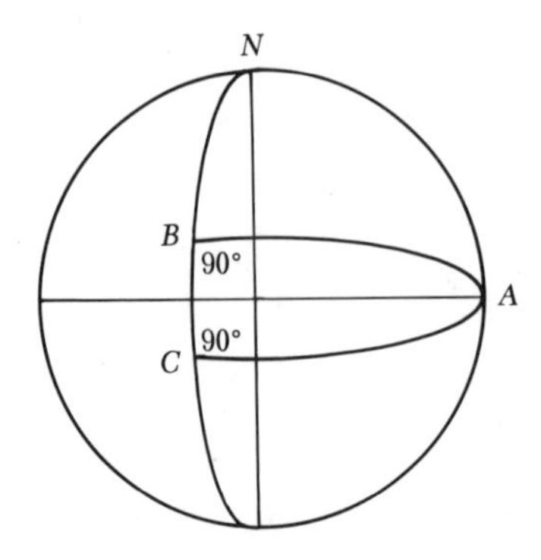

FIG. 3.28

An immediate consequence of this theorem is:

Theorem 3.12 *The sum of the angles of a quadrilateral is greater than* 360°.

Because the angle sum of a triangle differs in Euclidean, hyperbolic, and elliptic geometry, the angle sum theorem is often used to classify these various systems. Thus, *according as the angle sum of a triangle is greater than, equal to, or less than* 180°, *the geometry is, respectively, elliptic, parabolic (Euclidean), or hyperbolic.*

Exercise 3.8

1. Draw figures illustrating each of the following elliptic theorems and prove them:
   a. In an isosceles triangle, the sides opposite equal angles may be unequal.
   b. The sum of two sides of a triangle may be less than the third side.
2. Why is Euclid's Theorem 16 (Appendix A) not necessarily a theorem of elliptic geometry? *Hint:* See ([21] p. 130).

3.17 The Saccheri and Lambert quadrilaterals in elliptic geometry

Saccheri's historical birectangular isosceles quadrilateral $ABCD$ with right angles at A and B and equal sides AD and BC is illustrated in Fig. 3.29. As in hyperbolic geometry, the summit angles C and D are equal, and by virtue of Theorem 3.11 these same angles are obtuse. Thus other theorems are:

Theorem 3.13 *The summit angles of a Saccheri quadrilateral are equal and obtuse.*

Theorem 3.14 *In a Lambert quadrilateral ABCD in which* $\angle A = \angle B = \angle C =$ 90°, *the fourth angle D is obtuse.*

Theorem 3.15 *There are no squares in elliptic geometry.*

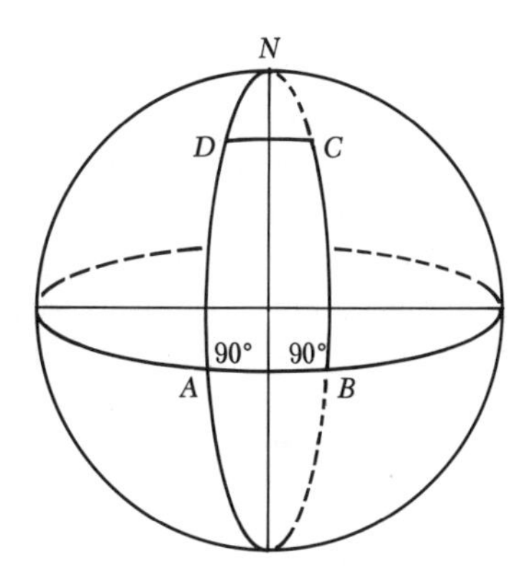

FIG. 3.29

Theorem 3.16 *Two similar triangles are congruent.*

Exercise 3.9

Which is greater, the base or the summit of the Saccheri quadrilateral in the elliptic plane? Prove your answer.

3.18 Elliptic geometry and the physical universe

The possibility of using elliptic theorems to describe the universe was explored by Gauss, who thought that he could settle once and for all the question of which one of the metric geometries *best* described the physical universe.

Since the theorem concerning the angle sum of a triangle differed in all three geometries—Euclidean, hyperbolic, and elliptic—Gauss decided to measure the angles of a large triangle. He needed a large triangle, since the angle sum of a triangle in all three geometries was known to be in close agreement for small regions, but differed radically in astronomically large domains. Gauss therefore stationed an observer on each of three fairly distant mountain peaks, with instructions to each one to measure the angle formed by light rays from his point of observation to each of the remaining observers.

When the data were collected, it was found that the angle sum was within 2 seconds of 180°, and so nothing was settled. Allowance had to be made for errors of measurement, for the fact that the triangle might not have been sufficiently large, and for the possibility that the light rays which formed the sides of his triangle were curved. However, the experiment, although inconclusive, did bring out the significant fact that all three geometries are in close agreement for sufficiently small regions. This situation closely parallels that in physics, where the systems of Newton and Einstein are in close agreement for small distances and velocities, but differ widely for astronomically large quantities.

Other experiments have been made, e.g., terrestrial experiments on two rays of light perpendicular to one plane. Results seem to indicate that these lines remain equidistant as expected in Euclidean geometry, but it is quite conceivable that in a sufficiently large domain they might diverge, as expected in hyperbolic geometry, or converge, as expected in elliptic geometry. Furthermore, conclusions based on measuring instruments which are in turn based on one or the other of these different geometries would hardly be convincing.

All this sums up to the conclusion that one cannot say which geometry best describes the physical universe.

Concluding remarks

The present introduction to non-Euclidean geometry has stopped at the threshold of vast collections of unexplored treasures. Early studies have mushroomed. In the calculus approach to geometry a study is made of the expression

$$ds^2 = \frac{dx_1^2 + dx_2^2 + dx_3^2}{1 + (K/4)(x_1^2 + x_2^2 + x_3^2)}$$

for arc length between two points (x_1,x_2,x_3) and $(x_1 + dx_1,x_2 + dx_2,x_3 + dx_3)$ of a space. The constant K is the curvature of the space: according as K is greater than, equal to, or less than zero, the space has constant positive curvature, zero curvature, or constant negative curvature.

For $K = 0$ the expression reduces to

$$ds^2 = dx_1^2 + dx_2^2 + dx_3^2$$

which will be recognized as the square of the differential of arc length in Euclidean space.

Riemann considered a geometry of space in which the curvature may change the character of the geometry from point to point. His later systems, which make use of the calculus, are usually referred to as "Riemannian" geometries. Although such systems are not to be considered here, it should be mentioned in passing that in Einstein's general theory of relativity, the geometry of space is based on a Riemannian geometry.

Many other investigations were made. The great physicist Helmholtz (1821–1894), for example, discussed the relation between the forms just given for arc length and the assumption of free mobility: "Spatial magnitudes can be moved from one place to another without distortion." Later the Norwegian mathematician S. Lie (1842–1899) linked Helmholtz's investigations with group theory and showed that there are three types of groups which characterize the three geometries: Euclidean, hyperbolic, and elliptic.

All the new theories have yielded deeper insights into the space concept. Whereas originally space meant the actual space of the universe, *today a space is simply a set of elements together with certain properties of these elements,* such as the distance between two of them. Topological space is considered in topology, a new development in geometry that has become one of the most potent forces in modern mathematics.

Suggestions for further reading

Bonola, R.: "Non-Euclidean Geometry."

Carslaw, H. S.: "The Elements of Non-Euclidean Geometry."

Cooley, H. R., D. Gans, M. Kline, and H. Wahlert: "Introduction to Mathematics," chap. 22.

Courant, R., and H. Robbins: "What is Mathematics?," pp. 214–227.

Coxeter, H. S. M.: "Non-Euclidean Geometry."

Eves, Howard: "A Survey of Geometry," vol. I. chap. 7.

Hilbert, D., and S. Cohn-Vossen: "Geometry and the Imagination."

Klein, Felix: "Elementary Geometry from an Advanced Standpoint," pp. 174–188.

Kline, Morris: "Mathematics in Western Culture," chap. 26.

Lieber, Hugh G., and Lillian R. Lieber: "Non-Euclidean Geometry."

Manning, Henry P.: "Non-Euclidean Geometry."

Moise, E.: "Elementary Geometry from an Advanced Viewpoint," chap. 9, pp. 24–27.

Prenowitz, Walter, and M. Jordan: "Basic Concepts of Geometry."

Wolfe, H. E.: "Introduction to Non-Euclidean Geometry."

Young, J. W.: "Lectures on Fundamental Concepts of Algebra and Geometry," pp. 14–35.

Young, J. W. A.: "Monographs on Topics of Modern Mathematics," pp. 93–151.

PART TWO

PURE, NONMETRIC PROJECTIVE GEOMETRY

4 euclidean forerunners of projective geometry

In marked contrast with Euclid's geometry, which is concerned almost exclusively with such metric concepts as lengths, angles, areas, and volumes, there was developed about 150 years ago a new science called projective geometry. In its purest form this science is free of the distance concept.

Roots of the new geometry are found in ancient theorems dealing with nonmetric properties of figures, i.e., theorems concerning points all lying on the same line, or lines all passing through the same point, but the real significance of such theorems was not recognized until hundreds of years later. They are now seen to be in reality projective theorems in a Euclidean setting.

An initial study of these Euclidean forerunners of projective geometry will serve a double purpose. It will bring to the attention of the reader some Euclidean theory of more than passing interest, and at the same time it will show how a new science is born.

After projective geometry has been developed as a logical system in its own right, quite independent of the early science, Euclidean and projective geometries will be linked by certain advanced viewpoints. Such a plan closely follows the historical and painstakingly slow development of the new science.

Two classical theorems which will introduce nonmetric theory in a metric setting are those of Menelaus and Ceva. The first of these theorems was discovered by Menelaus of Alexandria about 100 B.C. and was rediscovered about 1,800 years later by Ceva, an Italian engineer and mathematician. In 1678, Ceva published both Menelaus' theorem and a very closely connected one of his own, named, after him, Ceva's theorem.

Reference theorems of Euclidean geometry upon which proofs of theorems depend will be found in Appendix A.

SOME CLASSICAL THEOREMS

4.1 Menelaus' theorem

Menelaus' theorem is concerned with the collinearity of points, one on each side of a triangle. Before considering the theorem, recall two definitions:

> *Two points A, B on a directed line determine a line segment whose length AB is positive if the direction from A to B agrees with the positive direction on the line and negative if the direction from A to B is in the contrary direction.*

Thus

$$AB = -BA$$

A point P divides a directed line segment AB into two line segments having the ratio r, where

$$r = \frac{AP}{PB}$$

The ratio r is therefore positive or negative according as P is an internal or an external point of division, or in other words, according as P is between A and B or B is between A and P. For example, point P of Fig. 4.1 divides side AB of triangle ABC into a negative ratio, whereas point Q divides side AC into a positive ratio.

A statement and proof of Menelaus' theorem follows:

Theorem 4.1 *A necessary and sufficient condition that three points P, Q, R on the respective sides or extensions of the sides BC, CA, AB of the triangle ABC be collinear is that*

$$\frac{AR}{RB} \cdot \frac{BP}{PC} \cdot \frac{CQ}{QA} = -1 \tag{1}$$

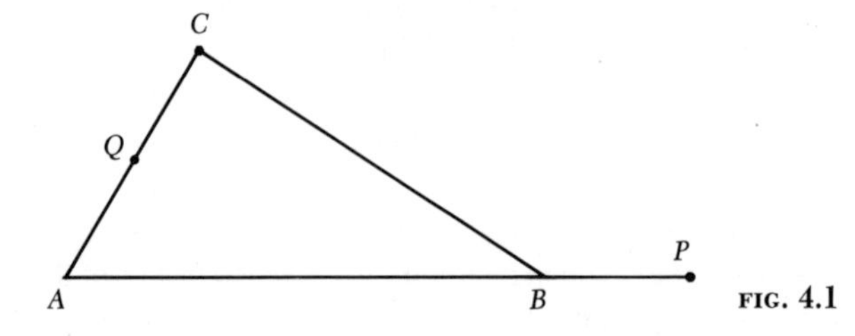

FIG. 4.1

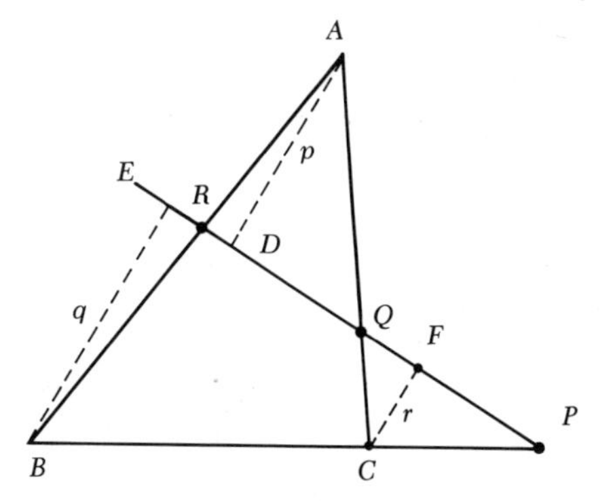

FIG. 4.2

The necessity of this condition will be established by showing that, when the points P, Q, R are collinear, condition (1) is satisfied.

Let p, q, r (Fig. 4.2) denote the lengths of the respective perpendiculars AD, BE, CF to the line PQR, called a transversal of the triangle ABC. Suppose first that none of the points P, Q, R coincides with a vertex. Then, since triangle ADR is similar to triangle BER, triangle CFP to triangle BEP, and triangle CFQ to triangle ADQ, and since only one of the points P, Q, R, say P, is an external point of division (Axiom 2.4, Appendix B),

$$\frac{AR}{RB} = \frac{p}{q} \qquad \frac{CQ}{QA} = \frac{r}{p} \qquad \frac{BP}{PC} = -\frac{q}{r}$$

Hence

$$\frac{AR}{RB} \cdot \frac{BP}{PC} \cdot \frac{CQ}{QA} = -\frac{p}{q}\,\frac{q}{r}\,\frac{r}{p} = -1$$

and the necessity of condition (1) has been proved.

The sufficiency of the given condition will be established by showing that from (1) follows the collinearity of the points P, Q, R.

Let a line through two of the points Q, R meet side BC in a point P'. Then from the proof just given

$$\frac{AR}{RB} \cdot \frac{BP'}{P'C} \cdot \frac{CQ}{QA} = -1$$

which, when divided by (1), gives

$$\frac{BP'}{P'C} = \frac{BP}{PC}$$

Hence

$$\frac{(BP' + P'C)(=BC)}{P'C} = \frac{(BP + PC)(=BC)}{PC}$$

from which it follows that $P'C = PC$. But points P' and P are both on BC extended. Therefore P' coincides with P, and the three points P, Q, R are collinear. Thus the theorem is proved when no point coincides with a vertex.

When a transversal passes through a vertex, say C, the points P and Q coincide, $PC = CQ = 0$, and condition (1) is meaningless. But then it follows that

$$AR \cdot BP \cdot CQ = RB \cdot PC \cdot QA$$

since each side has the value zero. This last statement illustrates, therefore, *the alternative form of Menelaus' theorem:*

> *If three points, taken on the three sides of a triangle, determine on these sides six segments such that the product of three nonconsecutive segments is equal to the product of the other three, the three points are collinear, and conversely.*

4.2 Applications

It is now possible to bring a number of apparently isolated theorems of elementary geometry into one general class, i.e., those theorems dealing with the collinearity of a triple of points, one on each side of a triangle.

One general method for handling all such theorems is provided by Menelaus' theorem. The method and the great unifying power of this classical theorem are illustrated next.

Theorem 4.2 *The internal bisectors of two angles of a triangle and the external bisector of the third angle meet their respective opposite sides in three collinear points.*

proof: In the triangle ABC (Fig. 4.3) let a, b, c denote lengths of sides opposite the respective vertices A, B, C, and let the external bisector t'_a of angle A and the two internal bisectors t_b, t_c of angles B and C meet their opposite sides in the respective points P, Q, R. Then from Theorem 59 of Appendix A it follows that

$$\frac{AR}{RB} = \frac{b}{a} \qquad \frac{CQ}{QA} = \frac{a}{c} \qquad \frac{BP}{PC} = -\frac{c}{b}$$

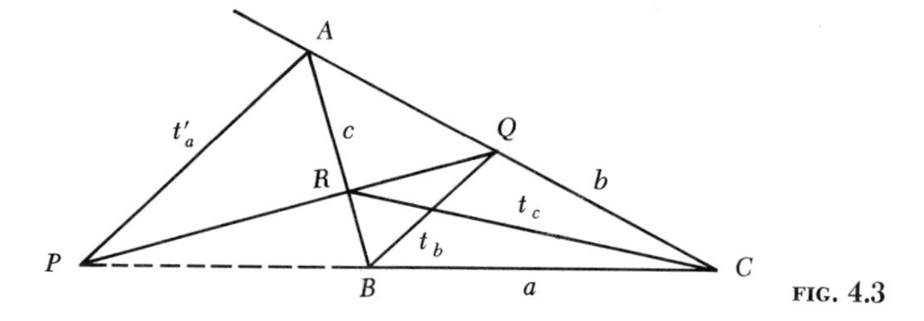

FIG. 4.3

and hence

$$\frac{AR}{RB}\cdot\frac{CQ}{QA}\cdot\frac{BP}{PC} = -\frac{a}{c}\frac{b}{a}\frac{c}{b} = -1$$

By Menelaus' theorem, the points P, Q, R are therefore collinear, and the theorem is proved.

Theorem 4.3 *Tangents to the circumcircle of a triangle at its vertices intersect the opposite sides in three collinear points.*

proof: Let the tangents to the circumcircle of triangle ABC at vertices A, B, C (Fig. 4.4) cut the sides BC, CA, AB at the points P, Q, R, respectively. Then in the triangles ABP and CAP, $\angle P$ is common, and $\angle PAB = \angle C$ (Theorems 57 and 55, Appendix A). They are therefore similar and

$$\frac{AB}{CA} = \frac{BP}{AP} = \frac{PA}{PC} \tag{1}$$

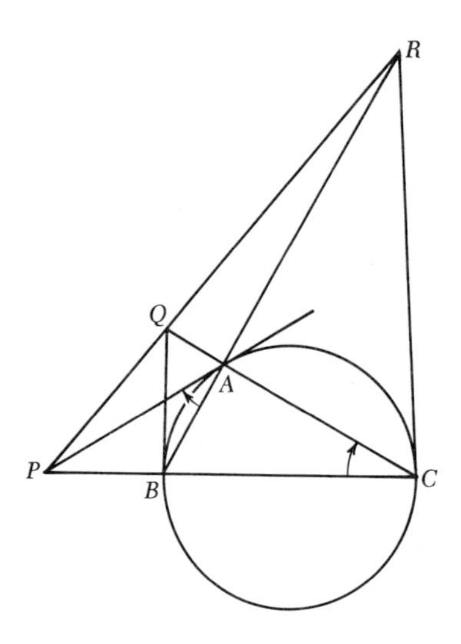

FIG. 4.4

Also, from the similarity of triangles ABQ and BCQ,

$$\frac{BC}{AB} = \frac{CQ}{BQ} = \frac{QB}{QA} \tag{2}$$

and from the similarity of triangles ACR and BCR,

$$\frac{AC}{CB} = \frac{CR}{BR} = \frac{RA}{RC} \tag{3}$$

Hence, multiplying (1), (2), and (3),

$$\left(\frac{AB}{CA} \cdot \frac{BC}{AB} \cdot \frac{AC}{CB}\right)^2 = \left|\frac{BP}{AP} \cdot \frac{PA}{PC} \cdot \frac{CQ}{BQ} \cdot \frac{QB}{QA} \cdot \frac{CR}{BR} \cdot \frac{RA}{RC}\right|$$

$$= \left|\frac{BP}{PC} \cdot \frac{CQ}{QA} \cdot \frac{RA}{BR}\right|^2$$

or, since P, Q, R are external points of division,

$$\left(\frac{BP}{PC} \cdot \frac{CQ}{QA} \cdot \frac{AR}{AB}\right) = -1$$

From Menelaus' theorem it then follows that the points P, Q, R are collinear, and the theorem is proved.

Exercise 4.1

Using Menelaus' theorem, prove each of the following theorems:

1 The external bisectors of a triangle meet the opposite sides in three collinear points.

2 If on the sides AB, AC of the triangle ABC are taken two equal segments AE and AF, the median issued from A divides the segment EF into the ratio of these sides.

3 A', B', C' are the midpoints of the sides BC, CA, AB of the triangle ABC. If line AA' meets the line $B'C'$ in point P and if the line CP meets AB and Q, then $AQ/AB = \frac{1}{3}$.

4 In the triangle ABC whose medians meet in the point G, AG is produced to a point P so that $GP = AG$. Then the parallels through P to CA, AB, and BC meet BC, CA, AB in three collinear points.

5 The incircle of the triangle ABC touches sides BC, CA, AB at the points X, Y, Z respectively. YZ is produced to meet BC in K; then

$$\frac{BX}{CX} = \frac{BK}{CK}$$

6 *Pappus' theorem:* "If A, B, C are any three distinct points on a line L, and A', B', C' any three distinct points on a second line L' of the plane, the three points

$$A'' = \binom{BC'}{B'C} \qquad B'' = \binom{CA'}{C'A} \qquad C'' = \binom{AB'}{A'B}$$

are collinear." (This theorem was discovered by Pappus of Alexandria about A.D. 300.)

4.3 Ceva's theorem

Ceva's theorem, which deals with the concurrency of triads, or triples of lines, one through each vertex of a triangle, is stated next and proved easily with the aid of Menelaus' theorem.

Theorem 4.4 *A necessary and sufficient condition that the three lines which join the points L, M, N on the respective sides BC, CA, AB of the triangle ABC to the opposite vertices be concurrent is that*

$$\frac{AM}{MC} \cdot \frac{CL}{LB} \cdot \frac{BN}{NA} = 1 \tag{1}$$

The necessity of this condition is shown first. Let each of the points L, M, N be distinct from a vertex of the triangle, and let P be the point of meeting of the lines AL, BM, and CN (Fig. 4.5). Then, in the triangle BNC cut by the transversal AL,

$$\frac{BA}{AN} \cdot \frac{NP}{PC} \cdot \frac{CL}{LB} = -1 \qquad \text{(Theorem 4.1)}$$

and similarly in the triangle ACN cut by the transversal BM,

$$\frac{AM}{MC} \cdot \frac{CP}{PN} \cdot \frac{NB}{BA} = -1$$

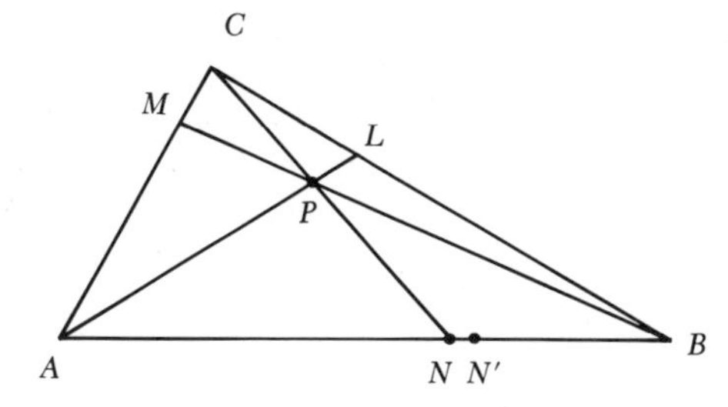

FIG. 4.5

Hence, multiplying these last two equations and simplifying,

$$\frac{AM}{MC} \cdot \frac{CL}{LB} \cdot \frac{BN}{NA} = 1 \tag{2}$$

Thus the necessity of (1) has been proved.

To show the sufficiency of this condition, let two of the three lines AL, BM, and CN, say BM and AL, intersect in a point P, and let the line CP meet AB at a point N'. Then

$$\frac{AM}{MC} \cdot \frac{CL}{LB} \cdot \frac{BN}{N'A} = 1 \qquad \text{(Why?)}$$

which, when divided by (1) and simplified, gives

$$\frac{BN'}{N'A} = \frac{BN'}{NA} \tag{3}$$

from which it follows that points N' and N coincide. The lines AL, BM, and CN are therefore concurrent, and the sufficiency of condition (1) has been established.

If any of the points L, M, N is a vertex, (1) is meaningless, but then

$$AM \cdot CL \cdot BN = MC \cdot LB \cdot NA$$

since each side of this equation has the value zero, and there is thus illustrated the *alternative form of Ceva's theorem:*

> *If three points taken on the three sides of a triangle divide these sides into six segments so that the product of three segments having no common end is equal to the product of the remaining three, the lines joining the three points to the opposite vertices of the triangle are concurrent, and conversely.*

4.4 Applications

Ceva's theorem, like Menelaus', unifies a number of elementary theorems dealing with the concurrency of triads of lines, each line being through a different vertex of a triangle, such as the altitudes or medians of a triangle, or the bisectors of its angles. In elementary geometry, these theorems are usually given separate treatments. Ceva's theorem, on the other hand, provides a standard method for handling all of them. The method is illustrated in what follows.

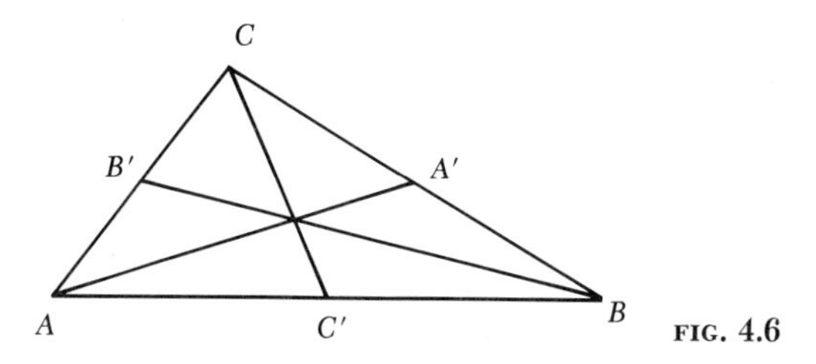

FIG. 4.6

Theorem 4.5 *The medians of a triangle are concurrent.*

The proof is simple. If A', B', C' are the midpoints of the sides BC, CA, AB of triangle ABC (Fig. 4.6),

$$\frac{BA'}{A'C} = \frac{CB'}{B'A} = \frac{AC'}{C'B} = 1$$

and hence

$$\frac{BA'}{A'C} \cdot \frac{CB'}{B'A} \cdot \frac{AC'}{C'B} = 1$$

It then follows from Ceva's theorem that the medians AA', BB', and CC' are concurrent.

The next theorem was proved in the early part of the nineteenth century by Joseph D. Gergonne, a French mathematician.

Theorem 4.6 *The lines joining the vertices of a triangle to the points of contact of the opposite sides with the incircle of the triangle are concurrent.*

proof: Let X, Y, Z (Fig. 4.7) be the points of contact of the incircle of triangle ABC with its sides BC, CA, AB. Then, from Theorem 53 of Appendix A,

$$AZ = AY$$
$$BZ = BX$$
$$CX = CY$$

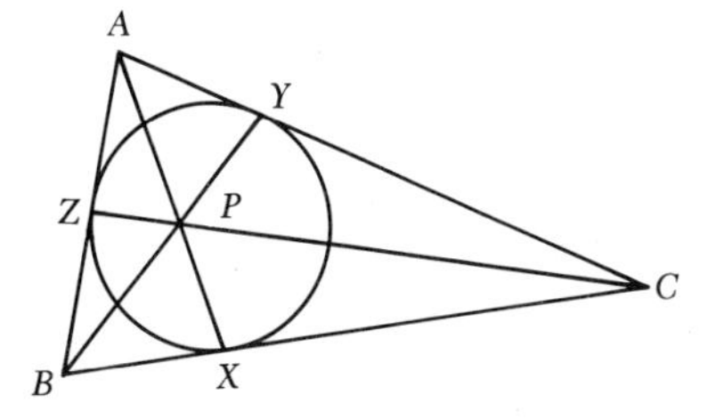

FIG. 4.7

and hence

$$\frac{AZ}{BZ} \cdot \frac{BX}{CX} \cdot \frac{CY}{AY} = 1$$

By Ceva's theorem, the lines AX, BY, CZ are therefore concurrent at a point P. *Point P is called the Gergonne point of the triangle.*

4.5 The pedal-triangle theorem

The next theorem depends on the following definition:

> *A line drawn from the vertex of a triangle to a point on the opposite side of the triangle is called a Cevian.*

The points L, M, N in Fig. 4.5 are the feet of the corresponding Cevians, and triangle LMN is called the *pedal triangle* of the point P. Also, P is called *the Ceva point of the triangle.*

Theorem 4.7 *If one of three concurrent Cevians in a given triangle is an altitude, this altitude bisects the corresponding angle of the pedal triangle.*

proof: Let AD be an altitude of the triangle ABC (Fig. 4.8), and let the three Cevians AD, BE, CF meet at a point P. To prove that AD bisects the angle FDE, draw a line through A parallel to BC and let this parallel intersect DE and DF in the respective points H and G. Then, triangle AFG is similar to BFD, and triangle AEH to triangle DEC. (Why?) Hence

$$\frac{AF}{FB} = \frac{AG}{BD} \qquad \text{and} \qquad \frac{CE}{EA} = \frac{DC}{AH} \tag{1}$$

But, by Ceva's theorem,

$$\frac{AF}{FB} \cdot \frac{BD}{DC} \cdot \frac{CE}{EA} = 1 \tag{2}$$

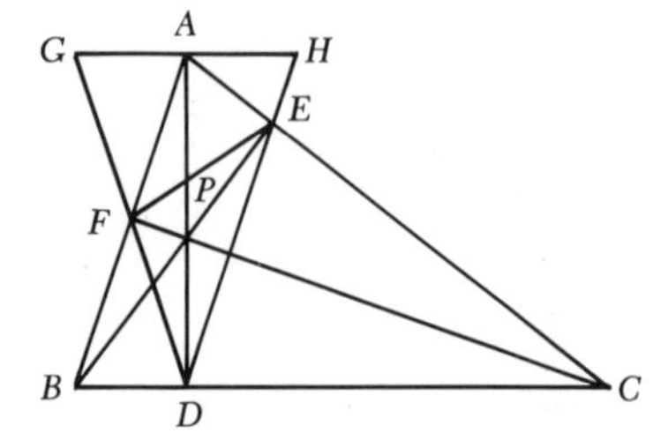

FIG. 4.8

Substitution of (1) in (2) then gives

$$\frac{AG}{BD} \cdot \frac{BD}{DC} \cdot \frac{DC}{AH} = 1$$

from which

$$AG = AH$$

Since the perpendicular from vertex D of triangle GDH bisects the base GH, it bisects also the angle FDE. Thus the theorem is proved.

4.6 Desargues' theorem

Desargues' famous theorem is stated and proved next.

Theorem 4.8 *If two triangles $A_1B_1C_1$ and $A_2B_2C_2$ are so situated that the three lines A_1A_2, B_1B_2, C_1C_2 joining corresponding vertices are concurrent, the pairs of corresponding sides will intersect in three collinear points.*

proof: If in Fig. 4.9 the lines A_1A_2, B_1B_2, C_1C_2 joining corresponding vertices of the two triangles $A_1B_1C_1$ and $A_2B_2C_2$ meet at a point O, and if A', B', C' are the respective points of intersection of corresponding sides B_1C_1, B_2C_2; A_1C_1, A_2C_2; A_1B_1, A_2B_2, the theorem will be proved by showing that the points A', B', C' are collinear. This is done by the use of Menelaus' theorem.

In the triangle A_2B_2O cut by the transversal A_1B_1C', it follows from Menelaus' theorem that

$$\frac{A_2C'}{C'B_2} \cdot \frac{B_2B_1}{B_1O} \cdot \frac{OA_1}{A_1A_2} = -1 \tag{1}$$

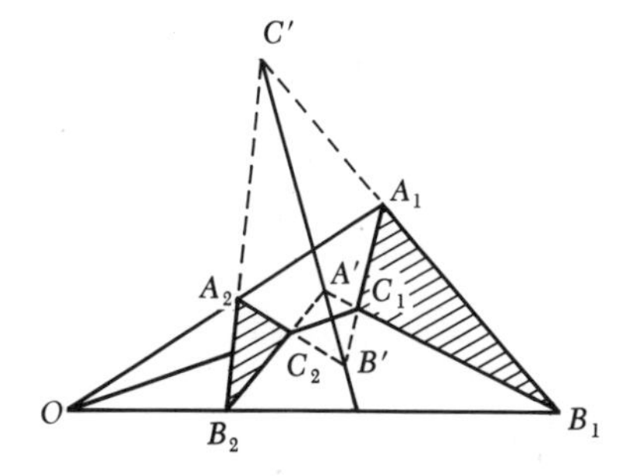

FIG. 4.9

Similarly, in the triangle B_2C_2O cut by the transversal B_1C_1A',

$$\frac{B_2A'}{A'C_2} \cdot \frac{C_2C_1}{C_1O} \cdot \frac{OB_1}{B_1B_2} = -1 \tag{2}$$

and, in the triangle A_2C_2O cut by the transversal A_1C_1B',

$$\frac{C_2B'}{B'A_2} \cdot \frac{A_2A_1}{A_1O} \cdot \frac{OC_1}{C_1C_2} = -1 \tag{3}$$

or, when (1), (2), and (3) are multiplied together and the result simplified,

$$\frac{A_2C'}{C'B_2} \cdot \frac{B_2A'}{A'C_2} \cdot \frac{C_2B'}{B'A_2} = -1$$

It then follows from Menelaus' theorem applied to triangle $A_2B_2C_2$ that the points A', B', C' are collinear, and the theorem is proved. It will appear later in an entirely different setting (see Sec. 5.4).

Remark The fact that many apparently isolated theorems of elementary geometry have been unified through the nonmetric theorems of Ceva and Menelaus suggests that a geometry in which the distance concept is *not* used may be significant. All the fascinating material of Chaps. 5 to 7 is nonmetric theory.

Exercise 4.2

1 Does Ceva's theorem hold if the point P (Fig. 4.5) is on a side of the triangle, or on a side produced? Explain.

2 Using Ceva's theorem, prove that the following triads of lines in a triangle are concurrent: (a) the altitudes, (b) the interior bisectors of the angles of the triangle, (c) the exterior bisectors of two angles and the interior bisector of the third angle.

3 Show that the centroid, i.e., the point of meeting of the medians of a triangle, is the only point which divides its Cevians into segments having equal ratios.

4 Show that, if one of three concurrent Cevians bisects the corresponding angle of the pedal triangle, it is an altitude of the given triangle.

5 If LMN (Fig. 4.5) is the pedal triangle of the Ceva point P for the triangle ABC, show that

$$\frac{PL}{AL} + \frac{PM}{BM} + \frac{PN}{CN} = 1 \qquad \text{and} \qquad \frac{CP}{PM} = \frac{CM}{AM} + \frac{CL}{LB}$$

6 Show that the circumcircle of the pedal triangle determined by three concurrent Cevians of a triangle ABC cuts the sides of the triangle in three other points such that the three Cevians formed by joining these points to the opposite vertices are concurrent.

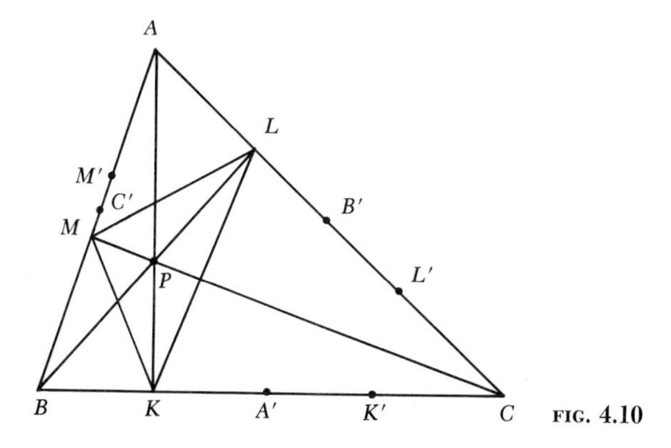

FIG. 4.10

7 The points P, Q, R, S lie on the sides AB, BC, CD, DA of the quadrilateral $ABCD$. Show that if PQ and RS meet on AC, then PS and QR meet on BD.

8 If (a) KLM (Fig. 4.10) is the pedal triangle of point P in triangle ABC, and if (b) A', B', C' are midpoints of sides a, b, c, and if (c) K' is symmetric to K with respect to A', L' to L with respect to B', M' to M with respect to C', show that the lines AK', BL', and CM' are concurrent.

PROBLEMS FOR SPECIAL STUDY

Exercise 4.3

1 The homothetic centers of two circles are the points which divide the line segment that is determined by their centers internally and externally into the ratio of their radii. Show that the six homothetic centers of three circles taken in pairs lie by three on four straight lines. *Hint:* See ([36] p. 23).

2 $ABCDEF$ (Fig. 4.11) is a simple hexagon inscribed in a circle, and the intersection of pairs of opposite sides are the points

$$L = \binom{AB}{DE} \qquad N = \binom{CD}{FA} \qquad M = \binom{BC}{EF}$$

Using Menelaus' theorems prove that the points L, M, N are collinear.

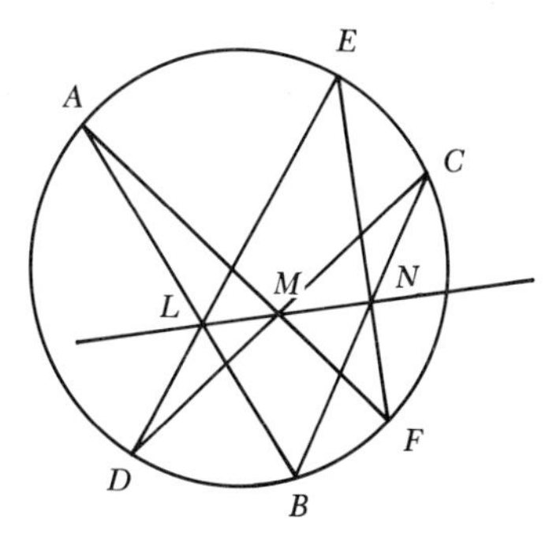

FIG. 4.11

3 Variable points E, F are taken on the sides AB, AC of triangle ABC so that $AF/FB + AE/EC$ is a constant. Show that the area of triangle BOC is constant, where O is the point of intersection of lines BE and CF.

4 A circle meets side BC of triangle ABC in points L, L'; CA in M, M'; and AB in N, N'. Show that if AL, BM, CN are concurrent, then AL', BM', CN' are also concurrent.

HARMONIC ELEMENTS AND CROSS RATIO

Another elementary theory to be treated later in an entirely different setting is concerned with a set of points and lines called *harmonic elements.* Their appearance in Euclidean geometry and their reappearance later in a basic role in modern, nonmetric developments of projective geometry will serve to emphasize some of the striking differences between the new and the old systems.

Harmonic points were discovered originally by Appolonius about 220 B.C. and were studied later by de la Hire (1640–1718) from a projective viewpoint.

It is especially interesting that the germ of the modern development was known to an ancient people. Modern man exploits the findings of the past.

4.7 Harmonic points. Basic theorems

Definition 1 *Four points A, B, C, D (Fig. 4.12) on a line are, by definition, harmonic if the points C and D divide the directed line segment AB internally and externally in the same ratio, and hence since line segments AD and DB are opposite in sign*

$$\frac{AC}{CB} = -\frac{AD}{DB}$$

For example, in the triangle APB (Fig. 4.12) the interior and exterior bisectors of angle P meet the opposite sides in the points C, D such that $AC/CB = -AD/DB$ (Theorem 59, Appendix A), and hence the points A, B, C, D are harmonic.

Definition 1 automatically separates the four points into the pairs (A,B) and (C,D) with one pair (A,B) forming the extremities of a line segment, and the other pair (C,D) its division points. Such a pairing is indicated by the symbol $H(AB,CD)$.

When

$$\frac{AC}{CB} = -\frac{AD}{DB} \qquad \frac{CA}{AD} = -\frac{CB}{BD}$$

and it is immaterial which pair of points represents extremities of the line segment, the symbols $H(AB,CD)$ and $H(CD,AB)$ represent the same set of points, and a second definition is:

Definition 2 *In the harmonic set of points H(AB,CD) either member of a pair AB (CD) is called the harmonic conjugate of the other member with respect to the remaining pair of points. Or, either division point C (D) is said to be the harmonic conjugate of the other division point D (C) with respect to the pair A, B.*

The name harmonic is well chosen. In music, when a set of strings of the same diameter and material are stretched to uniform tension and made to vibrate at one time, three strings produce harmony if their lengths form a harmonic progression.

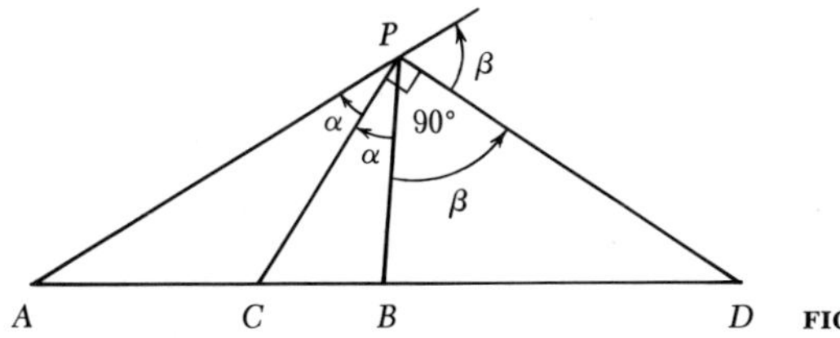

FIG. 4.12

Now, the lengths of the segments AB, AC, AD (Fig. 4.12) formed by the harmonic points $H(AB,CD)$ satisfy the condition

$$\frac{2}{AB} = \frac{1}{AC} + \frac{1}{AD}$$

and hence form a harmonic progression. This is stated and proved formally in the theorem:

Theorem 4.9 *In the harmonic set of points H(AB,CD) the lengths of the line segments AC, AB, AD form a harmonic progression, and conversely.*

proof: From definition 1

$$\frac{CB}{AB \cdot AC} = \frac{BD}{AB \cdot AC}$$

or, since $CB = AB - AC$ and $BD = AD - AB$,

$$\frac{AB - AC}{AB \cdot AC} = \frac{AD - AB}{AB \cdot AD}$$

which on simplifying and rearranging becomes

$$\frac{2}{AB} = \frac{1}{AC} + \frac{1}{AD}$$

and the segments AC, AB, AD form a harmonic progression. Since each of these steps is reversible, the theorem is proved.

A second elementary property of harmonic points is seen in the next theorem.

Theorem 4.10 *In the harmonic points H(AB,CD) the distance from the midpoint O of the line segment AB to each of its extremities is the mean proportional between the distances from O to the points C and D, and conversely.*

proof: Substitution of the relations

$$AC = OC - OA \qquad AD = OD - OA \qquad CB = OB - OC$$
$$DB = OB - OD \qquad OA = -OB$$

in the relation $AC/CB = -AD/DB$ for harmonic points $H(AB,CD)$ gives

$$\frac{OC - OA}{OB - OC} = \frac{OD - OA}{OD - OB}$$

or

$$2(OD \cdot OC + OA \cdot OB) = (OA + OB)(OD + OC) = 0$$

and hence

$$OC \cdot OD = \overline{OA}^2 = \overline{OB}^2$$

Since each step in the proof is reversible, the theorem is proved.

4.8 Applications

Some unusual and unexpected places in which harmonic points appear are shown in the next three theorems.

Theorem 4.11 *If two circles are orthogonal and if a diameter AB of one of them meets the other in the points C and D, the points A, B, C, D form the harmonic set H(AB,CD).*

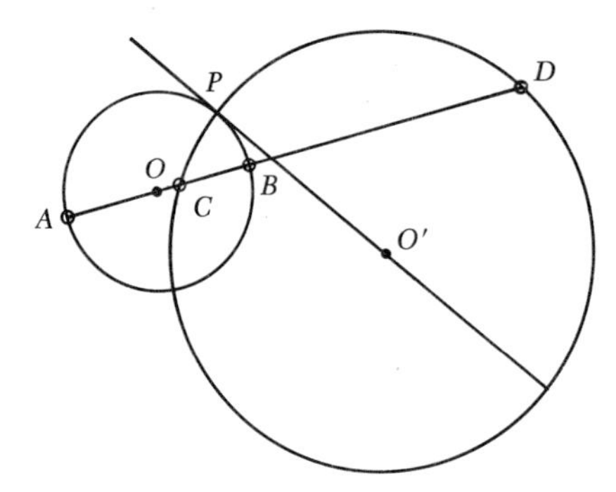

FIG. 4.13

proof: Let the given circles with centers O and O' (Fig. 4.13) intersect at point P. Then, from Theorem 61 of Appendix A, $\overline{OP}^2 = OC \cdot OD$, or since $OA = OP$, $\overline{OA}^2 = OC \cdot OD$, and by Theorem 4.10, the points A, B, C, D form the harmonic set $H(AB,CD)$.

Theorem 4.12 *If a chord BD of a circle is perpendicular to a diameter AC and lines from any point P of the circle to points B and D intersect AD in points B′ and C′, then the points A, B′, C′, D form the harmonic set H(AC′,B′D).*

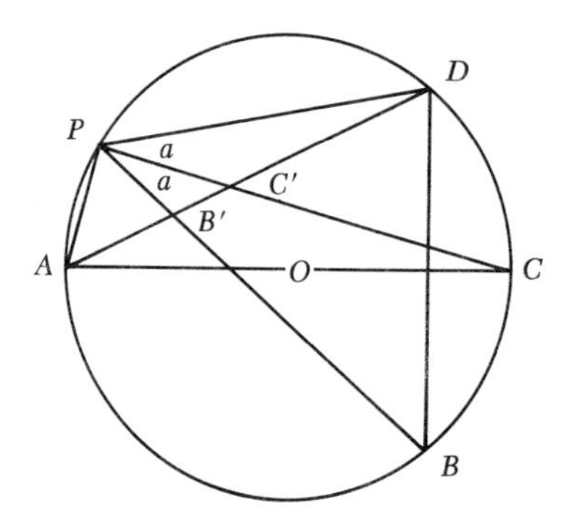

FIG. 4.14

proof: From the perpendicularity of the lines AC and BD follows the equality of arcs BC and CD (Fig. 4.14) making lines PC' and PA' the respective interior and exterior bisectors of angle BPD. Hence

$$\frac{AD}{DC'} = -\frac{AB'}{B'C'} \qquad \text{(Theorem 59, Appendix A)}$$

and from definition 1 of Sec. 4.1 the points A, B', C', D form the harmonic set $H(AC',B'D)$.

Theorem 4.13 *Concurrent lines through the vertices A, B, C of a triangle meet the opposite sides BC, CA, AB in the points L, M, N respectively. If the line MN meets the side BC in point L′, the points B, C, L, L′ are harmonic.*

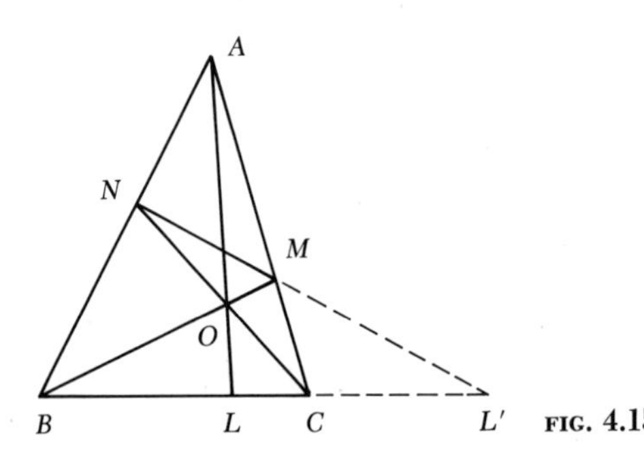

FIG. 4.15

proof: Let points and lines of Fig. 4.15 satisfy the conditions of the theorem. An application of Ceva's theorem (Theorem 4.4) to the concurrent lines AL, BM, CN gives

$$\frac{AN}{NB} \cdot \frac{BL}{LC} \cdot \frac{CM}{MA} = 1 \tag{1}$$

Also, since L', M, N are collinear, it follows from Menelaus' theorem that

$$\frac{AN}{NB} \cdot \frac{BL'}{L'C} \cdot \frac{CM}{MA} = -1 \tag{2}$$

Hence, dividing (1) by (2) and simplifying,

$$\frac{BL}{LC} = -\frac{BL'}{L'C}$$

and the points B, L, C, L' form the harmonic set $H(BC,LL')$. Thus the theorem is proved.

4.9 Harmonic point construction

From the proof of Theorem 4.13 there may be extracted a particularly simple method for constructing *with straightedge only* the harmonic conjugate of one of three given points.

If B, C, L (Fig. 4.15) are the three given points and the harmonic conjugate L' of L is to be constructed, simply connect the three given points

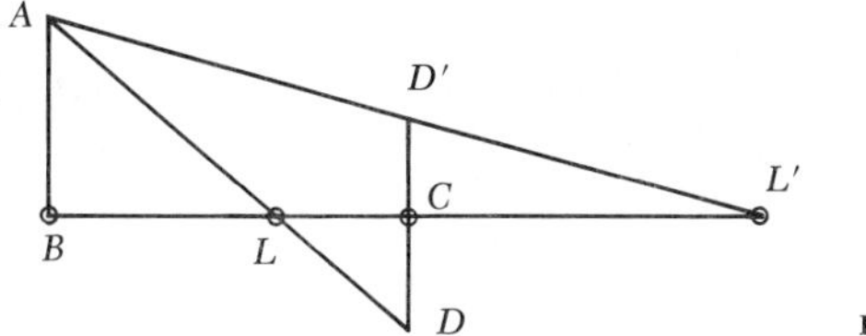

FIG. 4.16

to a point A not on their line, and through a point O on line AL draw lines BO and CO intersecting lines AC and AB in points M and N.

From the proof of Theorem 4.13, it then follows that the intersection of lines MN and BC is the desired point L'.

Contrast this simple straightedge construction with the older method which used not only the straightedge but a pair of dividers for marking off distances. Through points B and C (Fig. 4.16) draw the perpendiculars to line BC, and through L draw a line intersecting these perpendiculars in points A and D. On the perpendicular CD mark off point D' so that

$$CD = CD'$$

and draw AD'. The intersection of AD' with BC is the required point L'. For, from the theory of similar triangles,

$$\frac{BL'}{L'C} = \frac{AB}{CD} = \frac{BL}{LC}$$

or if line segments are directed,

$$\frac{BL}{LC} = -\frac{BL'}{L'C}$$

Exercise 4.4

1 Show that the lines joining any point on a circle to the vertices of an inscribed square are cut by a line L in a harmonic set of points.

2 Show that if A, B, C, D form the harmonic points $H(AB,CD)$ and if AL, AM are the arithmetic mean and the geometric mean of AC and AD respectively, then AM is a mean proportional to AL and AB.

3 If A, B, C, D and A', B', C', D' are two sets of harmonic points $H(AB,CD)$, $H(A'B',C'D')$ such that the lines AA', BB', CC', meet in a point O, prove that line DD' also passes through O.

4 The bisector of angle A of the triangle ABC meets the opposite side in the point P. Points Q and R are the feet of the perpendiculars from B and C to AP. Show that the set of points A, P, Q, R is harmonic.

5 The tangents at points P and Q to a circle intersect at point A, and its diameter produced passes through A. Show that A and Q are harmonically separated by the points in which their line is intersected by PB and PC.

6 Points L, M, N are the midpoints of the sides BC, CA, AB of the triangle ABC. Lines LA and LN meet BM in points X and Y. Show that the set of points M, X, Y, B is harmonic.

7 A line through the midpoint A' of the side BC of the triangle ABC meets the side AB in point F, side AC in G, and the parallel through A to side CB in E. Show that the set of points A', E, F, G is harmonic.

8 Prove that the line through the points of contact of the incircle with two sides of a triangle cuts the third side in a point which forms a harmonic set of points with the point of contact and the other two vertices on this third side.

9 Three lines, one issuing from each vertex of a triangle, are concurrent. The harmonic conjugate of the intersection of each line with the opposite side with respect to the other two vertices is determined. Prove that these three conjugate points are collinear.

CROSS RATIO

4.10 Cross ratio of four points

The relation $AC/CB = -AD/DB$ between the harmonic points $H(AB,CD)$ may be rewritten in the form

$$\frac{AC/CB}{AD/DB} = -1$$

in which the left hand side is seen to be the quotient of two ratios and the right hand side to have the special value -1. It is proposed to investigate this double ratio for any four points of a line and thus generalize the notion of a harmonic set of points.

The cross ratio of four points is by definition the quantity given by the formula

$$\lambda = \frac{\text{ratio into which } C \text{ divides line segment } AB}{\text{ratio into which } D \text{ divides line segment } AB} = \frac{AC/CB}{AD/DB}$$

If the points are distinct, $\lambda \neq 0$; and since there is only one point which divides a line segment into a given ratio, $\lambda \neq 1$. Moreover, λ is a signed quantity. When one direction on the line of the points is chosen to be positive, λ will be positive if the division points C, D are both external (Fig. 4.17a) or both internal (Fig. 4.17b). If one of the points of division is internal and the other external (Fig. 4.17c), λ is negative. Reversing the positive direc-

$C\;A\;B\;D$	$A\;C\;D\;B$	$A\;C\;B\;D$
$\lambda > 0$	$\lambda < 0$	$\lambda > 0$
(a)	(b)	(c)

FIG. 4.17

tion on the line merely changes the sign of every term of the cross ratio, and hence its value is independent of the direction chosen. The definition of cross ratio automatically separates the four points into pairs, so that points A, B of one pair are extremities of a line segment, whereas the remaining points C, D are its division points. Such a grouping is indicated by the symbol $R(AB,CD)$, or simply (AB,CD). Thus, if λ is the numerical value of $R(AB,CD)$,

$$\lambda = R(AB,CD) = (AB,CD) = \frac{AC/CB}{AD/DB} = \frac{AC \cdot DB}{CB \cdot AD} \tag{1}$$

Since any one of four distinct points A, B, C, D may be paired with any of the remaining ones, and since each permutation of the four letters A, B, C, D determines a particular cross ratio, order is an essential part of the definition of cross ratio.

Does the possibility of arranging four letters in 24 different ways mean that there are 24 possible cross ratios for any given four points on a line? This question will be investigated with the aid of the following more convenient formula for cross ratio.

Let a, b, c, d represent the distances of the collinear points A, B, C, D from a reference point O of the line. Then

$$AC = c - a \qquad CB = b - c \qquad AD = d - a \qquad DB = b - d$$

and (1) reduces to

$$(AB,CD) = \frac{(c-a)(b-d)}{(b-c)(d-a)} = \lambda \tag{2}$$

4.11 The six distinct values of the cross ratios

It will now be shown that the 24 cross ratios represented by the 24 permutations of the four letters A, B, C, D group themselves into six sets of four letters each, with each cross ratio in a set having the same value.

First it is noted that the cross ratio (CD,AB) is, by Eq. (2) of Sec. 4.10,

$$(CD,AB) = \frac{(a-c)(d-b)}{(d-a)(b-c)} = \lambda \tag{1}$$

and hence

$$(AB,CD) = (CD,AB) = \lambda$$

Also,

$$(BA,DC) = (DC,BA) = \lambda = \frac{(d-b)}{(a-d)}\frac{(a-c)}{(c-b)}$$

Thus

$$(AB,CD) = (CD,AB) = (BA,DC) = (DC,BA) = \lambda$$

and proof has been given of the theorem:

Theorem 4.14 *The cross ratio (AB,CD) does not change its value if the pairs of elements are interchanged, or if the order of elements in both pairs is reversed.*

Consider next the effect of reversing the order of the elements in a single pair. Then

$$(BA,CD) = \frac{(c-b)(a-d)}{(a-c)(d-b)} = \frac{1}{(c-a)(b-d)/(b-c)(d-a)} = \frac{1}{\lambda} \qquad (2)$$

and from Theorem 4.14,

$$(BA,CD) = (CD,BA) = (AB,DC) = (DC,AB) = \frac{1}{\lambda}$$

and proof has been given of the theorem:

Theorem 4.15 *The reversal of the order of elements in one pair of a cross ratio (AB,CD) changes its value λ to the reciprocal value $1/\lambda$.*

The effect of splitting pairs is noted next. Splitting the pairs A, B and C, D gives

$$(AC,BD) = \frac{(b-a)(c-d)}{(c-b)(d-a)} = \frac{ac - bc - ad + bd}{(b-c)(d-a)}$$

$$= \frac{(b-c)(d-a) - (c-a)(b-d)}{(b-c)(d-a)} = 1 - \lambda \qquad (3)$$

and another splitting gives

$$(AD,BC) = \frac{(b-a)(d-c)}{(d-b)(c-a)} = \frac{ad+bd+ac-bc}{(b-d)(c-a)}$$
$$= 1 - \frac{(b-c)(d-a)}{(c-a)(b-d)} = 1 - \frac{1}{\lambda} = \frac{\lambda - 1}{\lambda} \tag{4}$$

The results of the splitting are summarized in the theorem:

Theorem 4.16 *The cross ratios (AC,BD) and (AD,BC), in which pairs of (AB,CD) are split, are respectively* $(AC,BD) = 1 - \lambda$ *and* $(AD,BC) = \lambda - 1/\lambda$.

Thus

$$(AC,BD) = (BD,AC) = (CA,DB) = (DB,CA) = 1 - \lambda$$

and

$$(AD,BC) = (BC,AD) = (CB,DA) = (DA,CB) = \frac{\lambda - 1}{\lambda}$$

The results established are contained in the next theorem

Theorem 4.17 *A cross ratio of four given collinear points has one of the following six values:*

$$\lambda \qquad \frac{1}{\lambda} \qquad 1-\lambda \qquad \frac{1}{1-\lambda} \qquad \frac{\lambda-1}{\lambda} \qquad \frac{\lambda}{\lambda-1}$$

These six quantities are all distinct unless the points are harmonic and $\lambda = -1$. Then

$$\frac{1}{\lambda} = -1 \qquad 1 - \lambda = 2 = \frac{\lambda - 1}{\lambda} \qquad \frac{1}{1-\lambda} = \frac{1}{2} = \frac{\lambda}{\lambda - 1}$$

and the six distinct cross ratios reduce to the three values

$$-1 \qquad 2 \qquad \tfrac{1}{2}$$

4.12 A cross-ratio construction problem

When three points A, B, C of a line L (Fig. 4.18) are given, the problem of constructing the fourth point D such that $(AB,CD) = \lambda$ is instructive. This problem is a generalization of that given in Sec. 4.9.

The construction makes use of Pasch's order axiom (Axiom 3.4, Appendix B) and the following customary sign convention: "If C, A', B' are distinct points of a line, the ratio CA'/CB' is greater than or less than zero according as A' and B' are or are not on the same side of C (Fig. 4.18*a* and *b*).

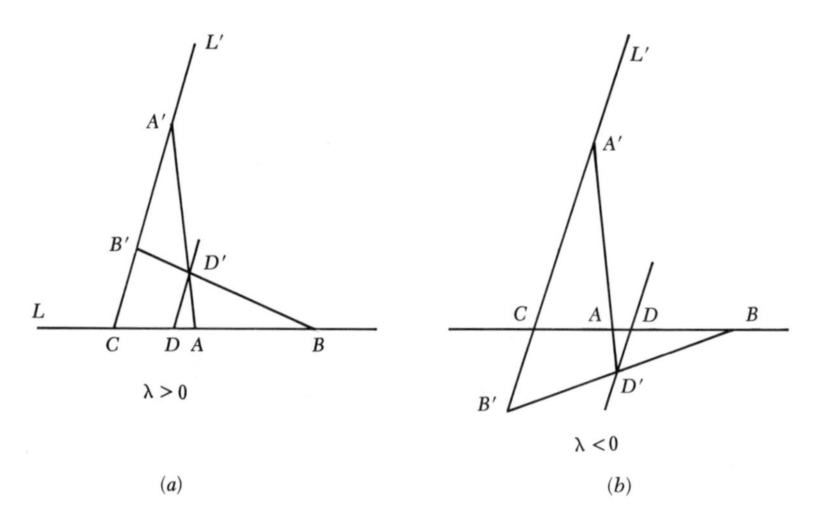

FIG. 4.18

Construction Problem Through C draw any line L', and on it take points A' and B' so that

$$\frac{CB'}{CA'} = \lambda$$

Draw AA' and BB'. If D' is their point of intersection, a parallel to CA' through D' meets L in the required point D.

proof: When all line segments are positive, it follows from the theory of similar triangles that

$$\frac{AC}{AD} = \frac{DD'}{CA'} \qquad \frac{CB}{BD} = \frac{CB'}{DD'}$$

and hence

$$\frac{AC}{AD}\cdot\frac{CB}{DB}=\frac{AC/CB}{AD/DB}=\frac{DD'}{CA'}\cdot\frac{CB'}{DD'}=\frac{CB'}{CA'}=\left|AB,CD\right|=\left|\lambda\right|$$

If line segments are directed, $CB'/CA' = (AB,CD) = \lambda$, for, from Pasch's axiom, D falls on AB extended when $\lambda > 0$ (Fig. 4.18*a*) and on line segment AB when $\lambda < 0$ (Fig. 4.18*b*).

Exercise 4.5

1. Which, if any, of the following cross ratios have the same value: (a) (AB,CD), (b) (AC,BD), (c) (BA,DC), (d) (BA,CD)? Give reasons for your answers.
2. State the theorems which justify the following equalities: (a) $(AC,BD) = (DB,CA)$, (b) $(BC,AD) = (DA,CB)$, (c) $(DC,AB) = (BA,CD)$.
3. If M is the midpoint of line segment AB, does the harmonic conjugate of M with respect to A and B exist? Explain.
4. If A, B, C are three distinct points on a line L, is there a point D on L distinct from A, B, or C such that $(AB,CD) = (BA,CD)$? Prove your answer.
5. If $(AB,CD) = (BA,CD) = \lambda$, give the value of λ.
6. If through the midpoint M of the chord AB of a circle two other chords CD and EF are drawn, and if DE and CF intersect AB in G and H, prove that M is also the midpoint of the segment GH.
7. If six points on a line correspond in pairs AA', BB', CC' and if $OA\cdot OA' = OB\cdot OB' = OC\cdot OC'$, where O is a point collinear with these six points, show that $(AA',BC) = (A'A,B'C')$.

4.13 An invariance property

Suppose now that four rays through a point O are cut by transversals L and L' in the respective sets of points A, B, C, D and A', B', C', D' (Fig. 4.19). Either set is then called *the projection of the other set from the center O of projection.*

FIG. 4.19

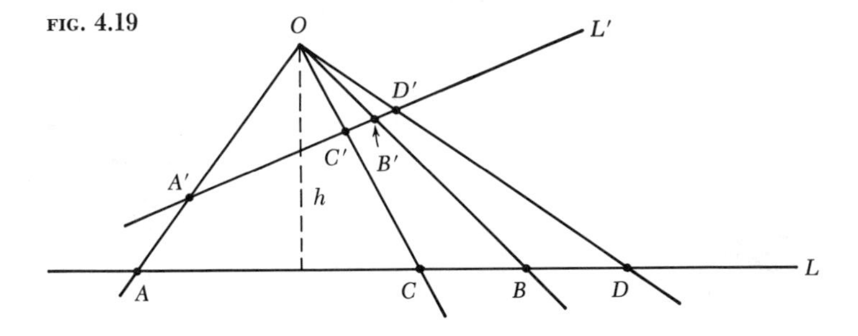

On projection, the lengths of line segments are seen to change, and in general, so will a ratio of line segments. In Fig. 4.19, $AC \neq A'C'$, and $AC/CB \neq A'C'/C'B'$. But a truly remarkable fact is noted now. There is a quantity which has not changed. It is the absolute value of the cross ratio (AB,CD) of the four points A, B, C, D. Only elementary geometry is needed to prove this fact. It is stated formally in the theorem:

Theorem 4.18 *The absolute value of the cross ratio of four points in the Euclidean plane is invariant on projection.*

proof: Let h denote the distance from the center O of projection to the line L containing the points A, B, C, D. Then, since the area of a triangle equals one-half the base by the altitude and is also equal to one-half the product of any two sides by the sine of the included angle,

$$\begin{aligned}
\text{Area } \triangle OAC &= \tfrac{1}{2}h \cdot AC = \tfrac{1}{2}OA \cdot OC \sin \angle AOC \\
\text{Area } \triangle OCB &= \tfrac{1}{2}h \cdot CB = \tfrac{1}{2}OB \cdot OC \sin \angle COB \\
\text{Area } \triangle OAD &= \tfrac{1}{2}h \cdot AD = \tfrac{1}{2}OA \cdot OD \sin \angle AOD \\
\text{Area } \triangle ODB &= \tfrac{1}{2}h \cdot DB = \tfrac{1}{2}OD \cdot OB \sin \angle DOB
\end{aligned}$$

from which it follows that

$$\begin{aligned}
\left|\frac{AC/CB}{AD/DB}\right| &= \left|\frac{OA \cdot OC \sin \angle AOC \cdot OD \cdot OB \sin \angle DOB}{OB \cdot OC \sin \angle COB \cdot OA \cdot OB \sin \angle AOD}\right| \\
&= \left|\frac{\sin \angle AOC \cdot \sin \angle DOB}{\sin \angle COB \cdot \sin \angle AOD}\right| = |\lambda|
\end{aligned}$$

and the quantity $|\lambda|$ depends only on the angles subtended at O by the segments AC, CB, DB, AD. Since these angles are the same for the segments $A'C'$, $C'B'$, $D'B'$, $A'D'$,

$$|AB,CD| = |A'B',C'D'|$$

as was to be proved.

Remark A proof that the cross ratio λ itself rather than its absolute value $|\lambda|$ is invariant on projection is best left for the more efficient methods of coordinate geometry.[1]

[1] See Sec. 10.4.

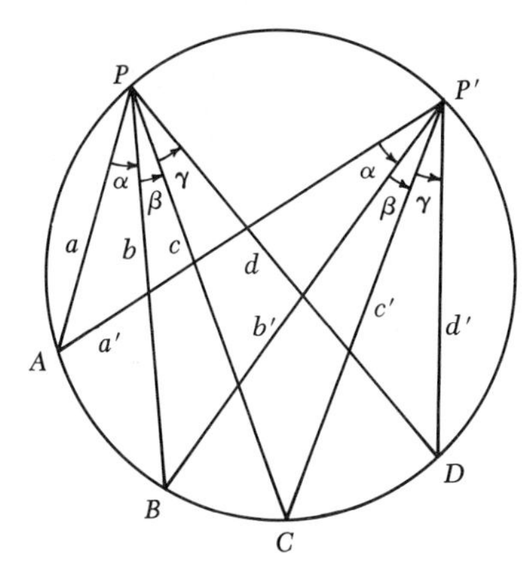

FIG. 4.20

PROBLEMS FOR SPECIAL STUDY

1 Let rays a, b, c through a point P and corresponding rays a', b', c' through a point P' form the equal acute angles, as shown in Fig. 4.20. If $A = (a,a')$, $B = (b,b')$, $C = (c,c')$, show that the five points P, P', A, B, C lie on a circle.

2 Prove that the cross ratio of the points of intersection of any four fixed tangents to a circle with a fifth tangent is the same for every position of the fifth tangent. *Hint:* Suppose the fixed tangents a, b, c, d at points A, B, C, D (Fig. 4.21) meet the fifth tangent at P in the points A', B', C', D' and any other tangents in the points A'', B'', C'', D''. Show that $|A'B',C'D'| = |A''B'',C''D''|$.

3 Using cross-ratio theory show that the points of intersection of opposite sides of a hexagon inscribed in a circle are collinear. *Hint:* See ([114] p. 108).

FIG. 4.21

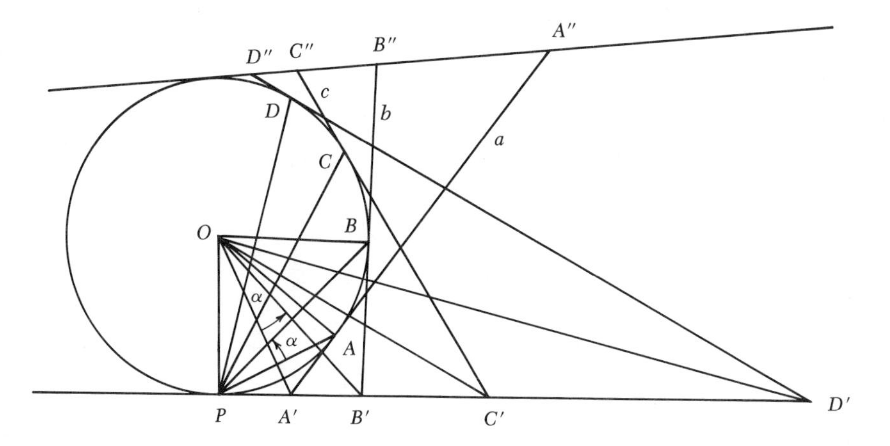

4 If A, B, C and A', B', C' are two sets of three points lying on the same straight line, and if there is a point D such that $(AB,CD) = (A'B',C'D)$, D is called a self-corresponding point in the two cross ratios. Can there be more than one self-corresponding point? Explain. *Hint:* See ([114] pp. 110–112).

Concluding remarks

This chapter has been devoted to a study of special Euclidean theory having both *simplifying* and *unifying* tendencies. After a statement and proof of the classical theorems of Menelaus and Ceva, Menelaus' theorem was used to provide a simple test for *collinearity of points,* and Ceva's theorem for *concurrency of lines.*

The chapter has also dealt with the notion of harmonic elements and its generalization, cross ratio. Again, the new concepts are seen to have both simplifying and unifying tendencies.

There is included also in this chapter a first introduction to an invariant in geometry. The notion is expanded in later sections of the text.[1]

Why all this material is called *projective geometry in a Euclidean setting* will be seen later when projective geometry is developed as an axiomatic system quite independent of Euclidean geometry.

Suggestions for further reading

Altshiller-Court, N.: "College Geometry."
Daus, Paul: "College Geometry."
Davis, David R.: "Modern College Geometry."
Johnson, R. Q.: "Modern Geometry."
Shively, L. S.: "Modern Geometry."
Taylor, E.E., and G. C. Bartoo: "An Introduction to College Geometry."

[1] See Chap. 10.

5 the formal approach to projective geometry

As so often happens in mathematics, it was a practical problem which gave the impetus needed for the formal development of projective geometry. The problem was literally forced on mathematicians by certain Renaissance artists, Leonardo da Vinci, Albrecht Dürer, and others, many of whom had noticed that in the process of depicting people, objects, or scenes on canvas, considerable distortion takes place. Lengths, angles, areas, and other metric properties may be changed, and still the identity of the original is unmistakable. The questioning artists wanted to know why this was true. The answer was obvious: Identifying or distinguishing characteristics, now called *invariants*, were retained.

Analyze for a moment the method by which a picture is painted. As the artist looks at an object, rays of light enter his eye, and when a transparent screen is placed between the eye and the object to be painted, these rays of light will meet the screen in a collection of points. It is this collection of points which the artist must draw if an observer is to receive the same impression from the painting as from the object itself.

The television is one of the more recent ways of reproducing elements by purely mechanical methods. An artist, on the other hand, must perform the process either by artistic skill alone or by the use of some formal laws. What are these laws?

The problem of the artist may be given a mathematical interpretation if physical concepts are replaced by their mathematical idealizations: the eye by a point O, the ray of light by a line, the screen by a plane M', and the object to be painted by a geometrical figure F lying in a plane M distinct from M' and not containing O.

Suppose the figure F consists of a circle in a plane M and a line intersecting it in two points A, B (Fig. 5.1). Then, if each point of the figure is joined to a point O not in M by a straight line, another plane M' will cut each of these lines in a point. The resulting collection F' of points is the image, or *projection*, of the original figure.

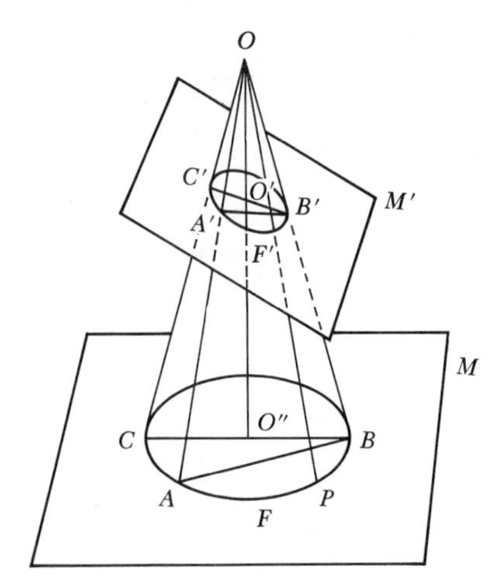

FIG. 5.1

The process just described has established a correspondence between points of a figure F and points of a figure F' in such a way that to each point of one figure corresponds one and only one point of the other.

If now F' is projected from another point O' onto a third plane, a new figure F'' results, and again a 1-1 correspondence is established between F' and F'' and hence between F and F''. This last correspondence is called a *projective correspondence.*

The special projective correspondence in which lines joining corresponding points meet at a single point is a *perspectivity.* Thus, all perspectivities are projective correspondences, but not all projective correspondences are perspectivities.

Figures F and F' are examined to see what properties of the original are retained in the projection. A straight line AB in F corresponds to a straight line $A'B'$ in F', and lines such as AB and AP which intersect have as their projections intersecting lines. Also, points on a line, such as C, O'', B, project into points C', O', B' which are also on a line, and so collinearity of points and concurrency of lines are invariant properties of a figure under projection. They are called *projective properties* in accordance with the definition:

> *A property of a figure which is preserved by every projection is a projective property.*

It is with projective properties of figures that projective geometry is primarily concerned. In fact, the profound and intensive search for the more deeply hidden of these projective properties led to some of the most unusual and remarkable discoveries in the new field.

The length of a line segment is not a projective property, but there is a transformation in which length is an invariant. It is the transformation of rigid motion, which, in a physical sense, is a movement of an object from one position in space to another, without in any way altering its shape or size. Such a transformation obviously preserves projective properties of figures, such as collinearity of points and concurrency of lines, but it does much more. *It preserves all magnitudes.*

> *A property which is preserved by all rigid motions is, by definition, a metric property.*

Distance, length, angle, and area are all metric notions, and theorems dealing with these concepts are called metric theorems. For example, the Pythagorean theorem (Theorem 47, Appendix A) is metric, as are most theorems of elementary geometry.

Figures, too, are classified as *metric* or *projective.* For instance, a triangle is a projective figure, since a triangle projects into a triangle, assuming, of course, that the center of projection is not in the plane of the triangle. An equilateral triangle, on the other hand, may project into a triangle having unequal sides; this means that the equilateral property may be lost on projection. An equilateral triangle is therefore a metric, not a projective, figure.

Again, since some circles project into ellipses, the property of being a circle is not projective. However, the general conic is projective since, as can be shown, a conic always projects into a conic.

The idea of freeing geometry of metric concepts, an idea which came only after the discovery of non-Euclidean geometry, was a result of the search for a science which would include all geometries, thus unifying the entire field. Recognition of such a possibility came when it was found that the essential character of Euclidean geometry and the basic non-Euclidean geometries was metric and that metric theorems differed in these various subfields.

The Englishman Arthur Cayley finally succeeded in showing that the Euclidean notion of distance could be defined in simple terms and was, in reality, only a particular case of a much more general projective definition. In 1859 he showed that both hyperbolic and elliptic non-Euclidean geometry could be defined as geometries associated with subgroups of the projective group. The proof emphasized the dominant role of projective geometry in the classification of geometries. Cayley's statement that "projective geometry is all geometry" is now classical.

Felix Klein later translated Cayley's algebraic methods into the language of pure geometry, and from that time on, projective geometry came to be recognized as the simplest type of geometry.

The invaluable contributions of Poncelet, Gergonne, von Staudt, Cayley, and Klein added to the works of other great mathematicians of the nineteenth and early twentieth centuries, such as those of Chasles, Steiner, Reye, and Cremona, brought to a grand climax the early investigations in this all-important field. All these men found in the new science a great aesthetic charm coupled with a most remarkable clarifying effect on geometry as a whole.

One final and important point is noted. In Cayley's projective definition of distance was found the solution of a number of problems in the ultramodern theory of special relativity. Once again, the mathematician had anticipated the needs of the scientist.

The plan in what follows is to develop projective geometry, *not as an extension of Euclid* but rather as a completely separate, formal system based on its own axioms concerning the undefined terms *point* and *line* and the two undefined relations *incidence* and *separation.*

One subset of these axioms—incidence axioms—deals with the property of a point being on a line; and a second subset—existence axioms—deals with, as the name suggests, the actual existence of points and lines. These two sets of axioms will be listed and discussed first. From these axioms, a number of important consequences will be obtained. Desargues' theorem is one of them.

The point is stressed at the start of this formal development of projective geometry that the distance concept does not appear in this chapter or in the two following chapters. The material presented is strictly nonmetric theory; nevertheless, its conclusions are many and far-reaching.

5.1 Incidence and existence axioms

The undefined elements are *points* $A, B, C, \ldots$ forming a set S and subsets $a, b, c, \ldots$ of them forming lines satisfying the following conditions:

Incidence Axioms

1. *If A and B are distinct points, there is at least one line on both A and B.*
2. *If A and B are distinct points, there is not more than one line on both A and B.*
3. *If A, B, C are points that are not all on the same line and if D, E are distinct points such that B, C, D are on a line and C, A, E are also on a line, there is a point F such that points A, B, F are on a line and points D, E, F are also on a line* (*Fig.* 5.2).

An immediate consequence of Axioms 1 and 2 is that the two points A, B determine a line AB, and a plane may now be defined:

Definition 1 *If A, B, C are* 3 *distinct noncollinear points of S, the plane p is the set of points lying on lines joining C to points of the line AB.*

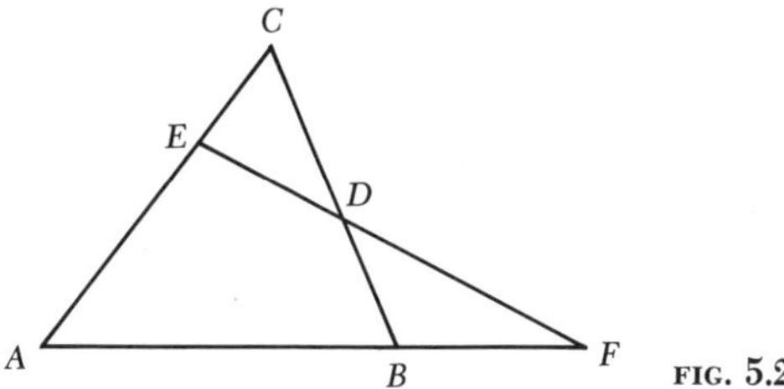

FIG. 5.2

From this definition and Axiom 3 it then follows that the plane p is equally well the set of points lying on lines joining B to A or C to B. The plane p is therefore determined by the three points A, B, C and two more definitions are:

Definition 2 *The totality of points on lines joining a point P not in the plane ABC to points of this plane is a projective* 3*-space, denoted* S_3.

Definition 3 *The figure determined by the three distance points A, B, C of a plane and the lines AB, BC, CA is a triangle. The points are its vertices and the lines its sides.*

This last definition permits the following convenient substitute for Axiom 3:

Axiom 3′ *If a line intersects two sides of a triangle, then it intersects the third side.*

The next set of axioms guarantees that the set S of points and the subset of lines are not empty.

Existence Axioms **4** *There exists at least one line.*
5 *There are at least three distinct points on every line.*
6 *Not all points are on the same line.*
7 *Not all points are on the same plane.*
8 *If S is a* 3*-space, every point is on S.*

When Axioms 7 and 8 are replaced by the single axiom:

7′ If p is a plane, every point is on p.

the resulting system is a plane geometry.

The material of the text is confined in the main to projective geometry of the *plane*, and it is assumed in what follows that the plane satisfies at least Axioms 1 to 6.

Some initial consequences of these axioms are obtained first.

5.2 Incidence properties in the plane

A first consequence of the initial axioms is stated formally in the theorem:

Theorem 5.1 *If two points of a line are on a given plane, every point of the line is on the plane.*

proof: If plane p is determined by a line m and a point C not on m, C and m are called *generating elements* of the plane.

If the given points D, E are on generating line m the theorem is an immediate consequence of the definition of a plane. If not, let lines CD and CE intersect m in points D' and E' (Fig. 5.3), forming with m the triangle $CD'E'$.

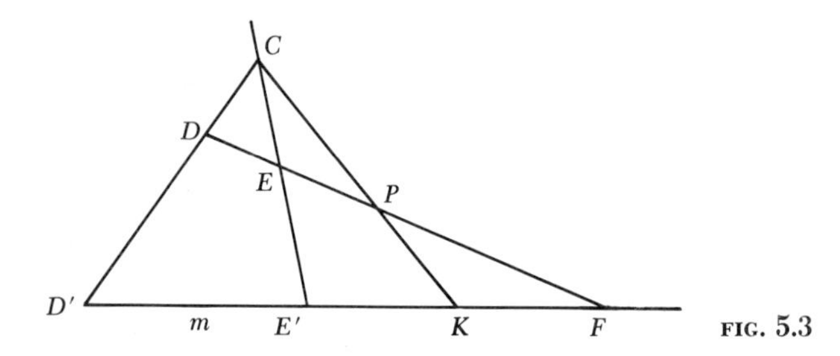

FIG. 5.3

By Axiom 3′ line DE cuts m in a point, say F. Then if P is any point of line DE, distinct from D and E, line CP cuts the two sides of $\triangle EE'F$; and again, by Axiom 3′, CP cuts the third side of this triangle in a point K. Point P is therefore on a line joining C to a point of the generating line m of the plane, and the theorem is proved.

The next theorem brings out a basic difference between the Euclidean and the projective plane. In the Euclidean plane, two lines do not meet if they are parallel. The situation in the projective plane is stated formally in the next theorem:

Theorem 5.2 *Any two distinct lines of the plane intersect.*

The proof makes use of the two lemmas:

Lemma 1 *A line L in the plane p always intersects its generating line m.*

Lemma 2 *Two lines a, b of plane p intersect if either one of them passes through the generating point C of the plane.*

To prove Lemma 1, it is noted first that if line L of plane p coincides with the generating line m, L intersects m in many points. Also, if line L passes through the generating point C, its intersection with m is an immediate consequence of the definition of a plane.

Then, if $L \neq m$ is a line of the plane and L is not through C and if A, B are two of its points, lines CA and CB will meet m in points A', B' (Fig. 5.4); and Axiom 3′ applied to $\triangle A'CB'$ gives a point Q in which L intersects m. Thus Lemma 1 has been proved.

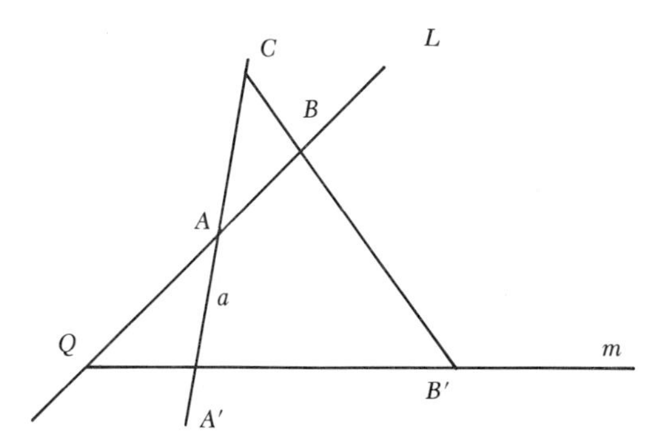

FIG. 5.4

To prove Lemma 2, let $a \neq b$ be the line through C intersecting m in A' (Fig. 5.5), and let the line joining C to any point B of line b intersect m in B'. From Lemma 1, line b intersects m in a point B'', and Axiom 3′ applied to $\triangle A'CB'$ gives a point A in which line b meets line a; and Lemma 2 has been proved.

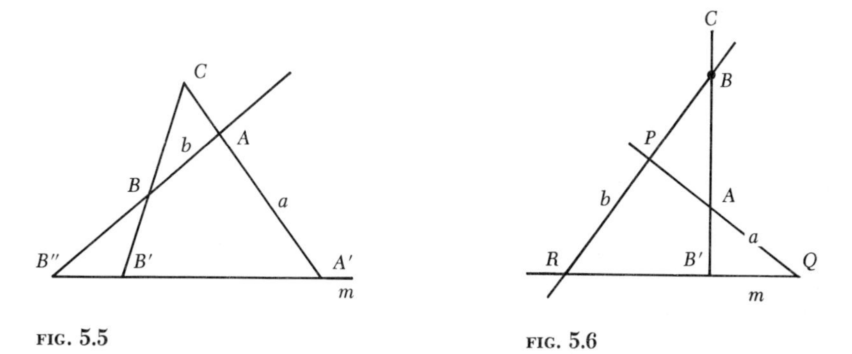

FIG. 5.5

FIG. 5.6

To prove Theorem 5.2 it is now only necessary to consider two lines a and b, neither of which passes through C or coincides with m (Fig. 5.6). Let Q, R be the respective points in which these lines meet m (Lemma 1). If B' is any point of line m, line $B'C$ meets each of the lines a and b in the respective points A and B (Lemma 2). If A and B coincide, $A = B$ is the desired point of meeting of lines a and b. Otherwise Axiom 3′ applied to $\triangle RBB'$ gives a point P in which, by Axiom 3′, line $AQ = a$ meets side $RB = b$, and thus Theorem 5.2 is proved.

Corollary *If one line of the plane contains n points, so does every line of the plane.*

This is so because, if line L of plane p contains n points and if L' is a line distinct from line L, there exists a point P not on either line; and to each point A of L corresponds the point A' on L' in which line PA meets L'.

5.3 Incidence properties in 3-space

Incidence properties for lines and planes of 3-space S_3 are given next. In the proofs that follow, plane p and a point P not on p denote generating elements of S_3.

Theorem 5.3 *If L is a line of S_3 not on the generating plane p of S_3, L meets p.*

To prove the theorem, let A, B be two points on L (Fig. 5.7), and let lines PA and PB meet p in points A', B'. By Axiom 3′, line AB meets line $A'B'$ in a point C which by Theorem 5.1 is in plane p.

Theorem 5.4 *A plane α of S_3 distinct from the generating plane p meets p in a line.*

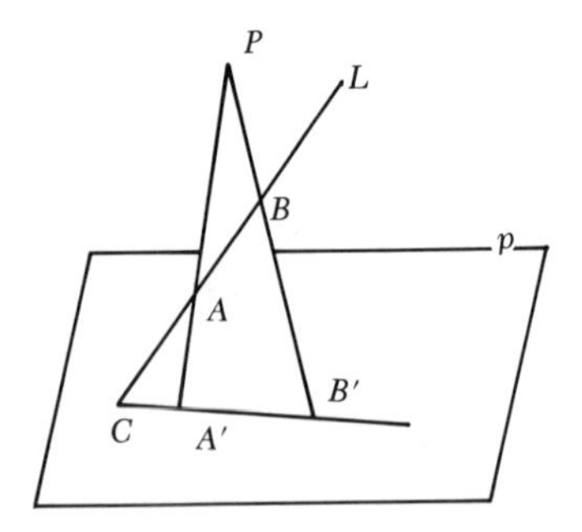

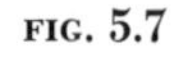
FIG. 5.7

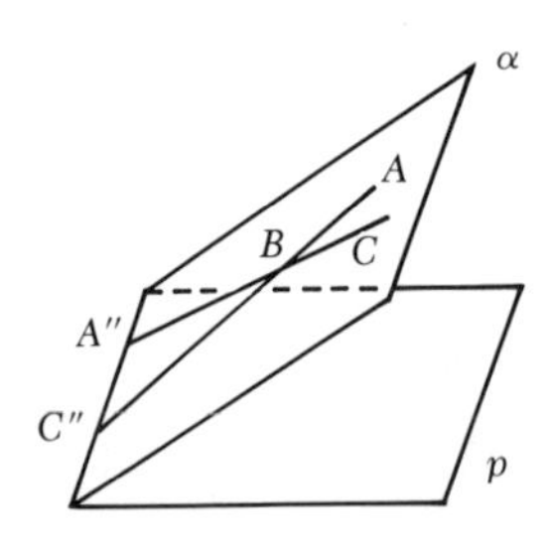

FIG. 5.8

proof: Let α be the plane determined by the intersecting lines AB and BC (Fig. 5.8), and let lines AB and BC meet plane p in the respective points C'' and A''. By Theorem 5.1, all points of line $A''B''$ are in both planes α and p. Since a point not on line $A''B''$ cannot lie in both planes α and p, the theorem is proved.

Theorem 5.5 *A plane α of S_3 distinct from the generating plane p and a line L of S_3 not on p always intersect.*

proof: Let plane α and generating plane p intersect in line m (Fig. 5.9), and let A be a point of α not on m. Then line L and point A determine a plane β distinct from α, meeting p in a line $n \neq m$. If B is the point of meeting of lines n and m, lines AB and L are both in plane β and hence meet at a point O lying in both α and β. Line L therefore meets α in O, and the theorem is proved.

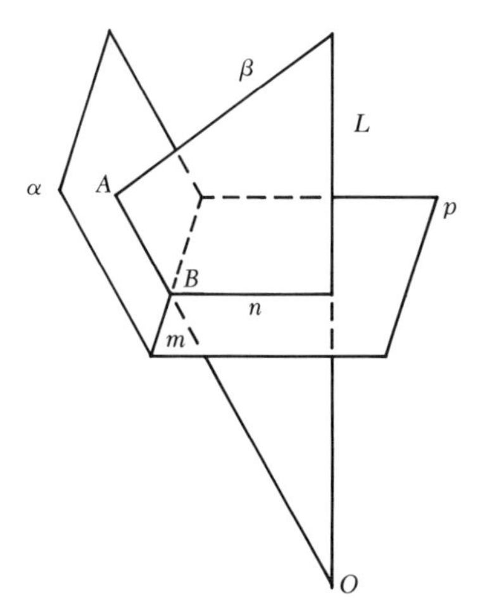

FIG. 5.9

Corollary 1 *Two distinct planes in a 3-space are on (intersect in) a line.*

Corollary 2 *Three distinct planes of S_3 not on a common line are on one and only one point.*

For, if three planes not on a common line intersect by pairs in lines a, b, and c, a point C common to two of these lines lies in all three planes. If there were a second point common to these three planes, they would all lie on a line, contrary to hypothesis.

Exercise 5.1

1 How does Axiom 3′ of Sec. 5.1 differ from Pasch's axiom (Sec. 2.10)?

2 Do points and lines of Fig. 2.4 satisfy Axioms 1 to 6 of Sec. 5.1? Explain.

3 Interpret the statement: "A line is on a point" to mean "A line passes through a point" and the statement: "Two lines are on a point" to mean "Two lines intersect in a point."

a. Using Axioms 1 to 6 of Sec. 5.1, prove each of the following theorems:

1′ If a and b are distinct lines of S, there is at least one point on both a and b.

2′ If a and b are distinct lines of S, there is not more than one point on both a and b.

3′ If a, b, c are lines of S not all on the same point and if d, e are distinct lines of S such that b, c, d are on a point and c, a, e are also on a point, there is a line f of S such that lines a, b, f are on a point and lines d, e, f are also on a point (Fig. 5.10).

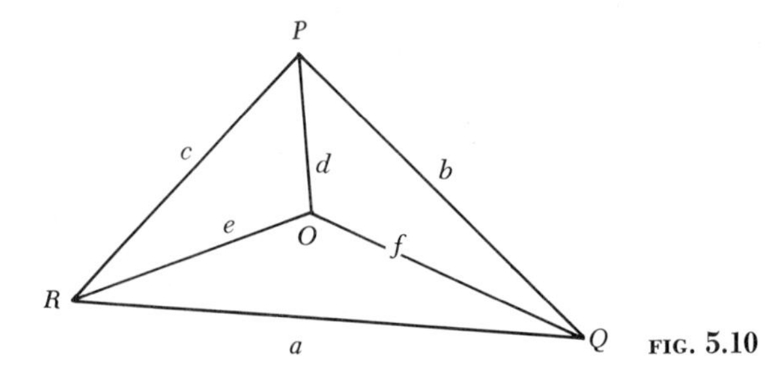

FIG. 5.10

4′ There exists at least one point in S.

5′ There are at least three distinct lines on every point of S.

6′ Not all lines of S are on the same point.

b. Show that these six theorems are obtained from Axioms 1 to 6 (Sec. 5.1) by simply interchanging the words "point" and "line."

4 On the basis of Axioms 1 to 8 of Sec. 5.1 and the assumption that there exists one line L containing exactly $(n + 1)$ points, prove that (a) every plane contains exactly $(n^2 + n + 1)$ points and (b) $(n^2 + n + 1)$ lines.

5.4 Desargues' theorem

So far no mention has been made of the very useful *distance* concept nor of number concepts and algebra. Parallel lines are missing and so are perpendicular ones. Yet, in many respects, the geometry resembles the familiar one of high school days, and one might reasonably ask what has been gained by this new approach. A proof of Desargues' famous triangle theorem will provide a partial answer to this question. This theorem, one of the imperishable treasures of mathematics, was published by Desargues somewhat informally in an appendix to a treatise on perspective written by a friend of his, Abraham Rose, a well-known engraver of the seventeenth century.

The earlier proof of Desargues' theorem given in Sec. 4.6 used metric concepts. They are superfluous. Stripping a proof of its metric covering, as will be done here, is like clearing the forest of the underbrush which hides the treasures underneath. Desargues' theorem holds in a simpler geometry than Euclid's. A statement and proof of the theorem follows.

Theorem 5.6 *If two triangles* $A_1B_1C_1$ *and* $A_2B_2C_2$ *are so situated that (a) the three* lines A_1A_2, B_1B_2, C_1C_2 *joining pairs of corresponding* vertices *are* concurrent *(on a point), then the three points of meeting of corresponding* sides *are* collinear *(on a line); (b) the three distinct* points *of meeting of corresponding* sides *are collinear, then the three* lines *joining corresponding* vertices *are* concurrent.

Theorems 5.6*a* and *b* will be recognized as converse theorems, and later they will be shown to be *dual* theorems.

The proof when the two triangles are in different planes is extremely simple. Let the lines A_1A_2, B_1B_2, and C_1C_2 (Fig. 5.11) joining pairs of corresponding vertices meet in the vertex O of the pyramid having these lines as edges, and let the lines A_1B_1 and A_2B_2 lying in the same plane intersect in a point, say C'. Similarly, let lines B_1C_1 and B_2C_2 meet at a point A', and lines A_1C_1 and A_2C_2 at a point B'. The points A', B', C' therefore lie in the plane of triangle $A_1B_1C_1$ and also in the plane of triangle $A_2B_2C_2$. By Corollary 1 of Theorem 5.5, these planes intersect in a line, and thus the theorem is proved for the case of triangles in different planes.

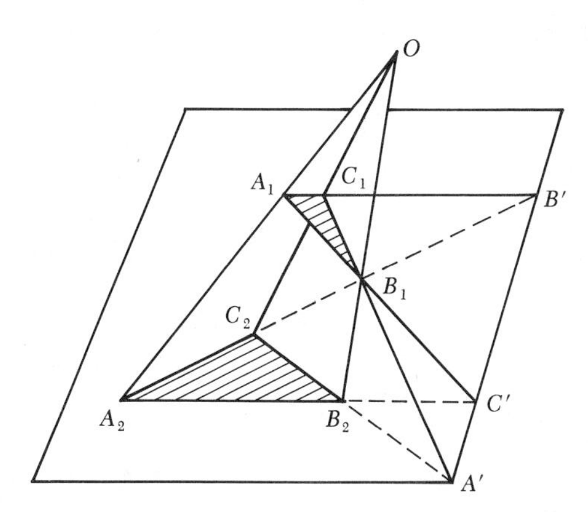

FIG. 5.11

The proof when the two triangles are in the same plane is considerably more complicated, but it can be reduced to the one just given. To do this, let P be the plane of the given triangles $A_1B_1C_1$ and $A_2B_2C_2$ (Fig. 5.12), and again let lines joining corresponding vertices meet at point O. Also, let a_i, b_i, c_i be the sides opposite the respective vertices A_i, B_i, C_i of these triangles, $i = 1, 2$. Choose any two points S_1 and S_2 collinear with O, both points being outside the plane P of the triangles.

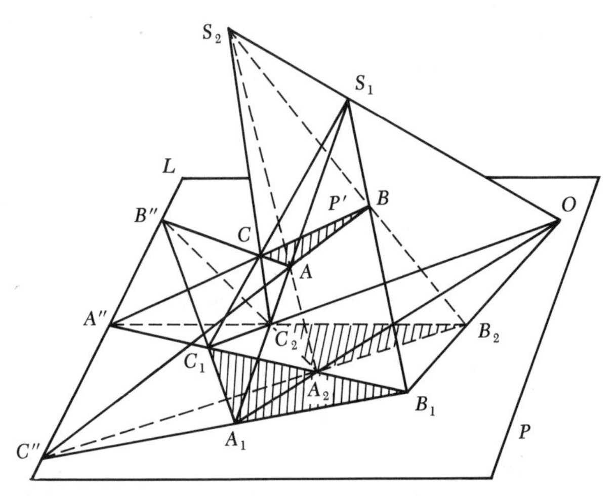

FIG. 5.12

Since the lines A_1A_2 and S_1S_2 intersect at point O, the lines S_1A_1 and S_2A_2 are coplanar and hence intersect at a point A. Similarly, the lines S_1B_1 and S_2B_2 intersect at a point B, and the lines S_1C_1 and S_2C_2 at a point C. Let P' be the plane of the triangle whose vertices are the points A, B, C, and let a, b, c be sides opposite the respective vertices A, B, C.

Triangles $A_1B_1C_1$ and ABC are in the distinct planes P and P', and by the case just proved, corresponding sides a_1 and a, b_1 and b, c_1 and c meet in the respective points A'', B'', C'' on the line L of intersection of these two planes P and P'.

Similarly, the triangles $A_2B_2C_2$ and ABC have corresponding sides a_2, a; b_2, b; c_2, c, which meet in points of this same line L. But—here is the key to the argument—there is only one point C'' in which line AB and hence lines A_1B_1 and A_2B_2 meet L. Similarly, AC meets L at the point B'' of meeting of lines A_1C_1 and A_2C_2, and BC meets line L at the point A'' of meeting of lines B_1C_1 and B_2C_2. The points A'', B'', C'' of meeting of corresponding sides in the two triangles therefore lie on a line L. Theorem 5.6a is therefore proved when the triangles lie in the same plane and the lines joining corresponding vertices are distinct.

A reversal of these arguments gives a proof of Theorem 5.6b.

The proof just given makes use of points in 3-space S_3. If the existence of such points is not postulated, the theorem cannot be proved on the basis of Axioms 1 to 6 alone ([100] pp. 126–129). A geometry in which Desargues' theorem does not hold is called non-Desarguesian ([57] pp. 74–99).

Some useful definitions at this point are:

Definition 1 *When two distinct triangles, such as* $A_1B_1C_1$ *and* $A_2B_2C_2$, *are so situated that lines joining pairs of corresponding vertices are con-*

current at a point O, the two triangles are said to be perspective from *O; and O is called the* center of perspectivity.

Definition 2 *When two triangles $A_1B_1C_1$ and $A_2B_2C_2$ are so situated that points of meeting of corresponding sides are on a line L, the triangles are said to be perspective from line L called the* axis of perspectivity *of the two triangles.*

An artist in drawing two perspective triangles probably knows from experience that the three points of meeting of pairs of corresponding sides will fall on the same line. If he has mastered Desargues' theorem, he will have the mathematician's answer to one of his early questions: "*Why* do these three points lie on a line?"

Exercise 5.2

1 Does Desargues' theorem hold (a) in the Euclidean plane if lines joining corresponding vertices are parallel, (b) in the affine geometry of Sec. 2.5?

2 Show by means of a diagram how 10 trees can be planted in 10 rows so that there are 3 and only 3 trees in each row.

3 Prove Theorem 5.6*b* if the two triangles are in different planes.

Prove each of the following theorems:

4 If three triangles have a common center of perspective, their three axes of perspectivity are concurrent.

5 If three triangles are perspective in pairs and if the pairs have a common axis of perspective, their centers of perspective are collinear.

6 If the triangles ABC and $A'B'C'$ are distinct and the lines AA', BB', CC' are concurrent, the six points of intersection of the pairs of lines AB, $A'B'$; BC, $B'C'$; CA, $C'A'$; $A'B$, AB'; $C'B$, CB'; $C'A$, CA' lie by three's on four straight lines.

5.5 Duality

Because two lines of the projective plane always meet, projective geometry possesses a distinctive characteristic not shared with the Euclidean system. It is the vitally important principle of duality illustrated first in the two statements:

Two lines are on a point.
Two points are on a line.

When, as in this case, an interchange of the words "point" and "line" in one statement gives the other, the two statements are called duals of each other. Other examples of dual statements are Axioms 1 to 6 of Sec. 5.1 and the corresponding Theorems 1′ to 6′ of Exercise 5.1, question 3.

The examples show how the introduction of a somewhat queer or awkward "on" language has made the process of dualizing a statement a more or less automatic matter of interchanging the words "point" and "line."

Figures too are said to be duals when they are graphic representations of a statement and its dual. Thus, two points A, B on a line and two lines a, b on a point, denoted (a,b), are dual figures. There is even a dual symbolism. The symbol (A,B) for a line on points A and B and the symbol (a,b) for a point on lines a and b are duals of each other.

With the aid of these preliminaries it is now possible to explain what is meant by the *principle of duality for the plane.*

Observe first that each of Axioms 1 to 6 of Sec. 5.1 has a dual which is a valid theorem (Exercise 5.1, question 3). Suppose then that a theorem dealing with the incidence of points and lines of the plane has been proved on the basis of these axioms. Then each of the statements in the proof has a dual, and the sequence of dual statements constitutes a proof of the dual theorem. Thus, if a theorem is deducible from a set S of statements and their duals, its dual is also deducible from the same set; and a formal statement of the principle of duality for the plane is, therefore:

> *Any properly worded valid statement concerning incidence of points and lines in the projective plane gives rise to a second valid statement obtained from the first by interchanging the words "point" and "line."*

Often the dual of a theorem is its converse, as is seen in Theorems 5.6a and b (Desargues' theorem); but special note is made that the principle of duality in the plane cannot be used to establish the truth or validity of Theorem 5.6a from Theorem 5.6b, for the proof of Theorem 5.6a made use of points and lines *not in the plane.*

There is a separate principle of duality for 3-space S_3 in which the dual elements are not point and line but *point* and *plane* with the word "line" left unchanged. For example, two dual statements in S_3 are:

> Three *points* not on the same line determine a plane.
> Three *planes* not on the same line determine a point.

Since the material of the text is confined in the main to the plane, the subject of space duality will not be pursued further.

Exercise 5.3

1 Does the principle of duality for the plane hold in Euclidean geometry? Why?

2 Form the duals of Theorems A and B below by filling in the blanks in Theorems A' and B', and draw figures illustrating each theorem and its dual.

Theorem A

If two vertices of a variable triangle move on fixed lines while the sides pass through three fixed collinear points, the third vertex will likewise move on a fixed line.

Theorem A′

If two ______ of a variable triangle pass through fixed ______ while the ______ lie on three fixed ______, the third ______ will likewise pass through a fixed ______.

Theorem B

If three points A, C, E are chosen at random on a fixed line p, and three points B, D, F at random on a fixed line q, the intersection of the lines AB with DE, BC with BF, CD with FA all lie on one line.

Theorem B′

If three ______ a, c, e are chosen at random on a fixed ______, and three ______ b, d, f at random on a fixed ______, the intersection of the points (a,b) with (d,e), (b,c) with (b,f), (cd) with (fa) all lie on one ______.

3 Draw and describe in words the dual of (a) a figure consisting of six lines a, b, c, d, e, f (Fig. 5.10) and four points P, Q, R, S such that on each point are three of the given lines and on each line are two of the given points; and (b) a figure consisting of four points P, Q, R, S (Fig. 5.13) such that on each point is a line and all lines are on one point O.

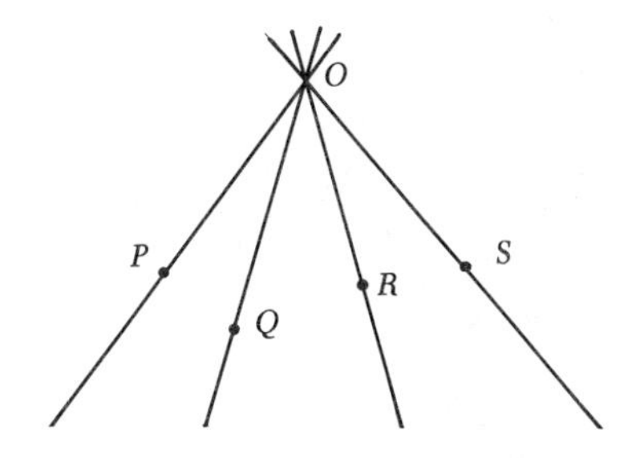

FIG. 5.13

5.6 The complete quadrilateral and its dual, the complete quadrangle

Two figures of basic importance in later developments are the *complete quadrilateral* and its *dual, the complete quadrangle.*

The complete quadrilateral (Fig. 5.14) consists of four *lines* (a_1, a_2, a_3, a_4) in a plane, no three of which are concurrent, and the six points P_1, P_2, Q_1, Q_2, R_1, R_2, in which pairs of these lines intersect. The lines are called *sides* and the points *vertices.* Any two vertices which do not lie on the same side are opposite, and a line joining a pair of opposite vertices is a *diagonal line.*

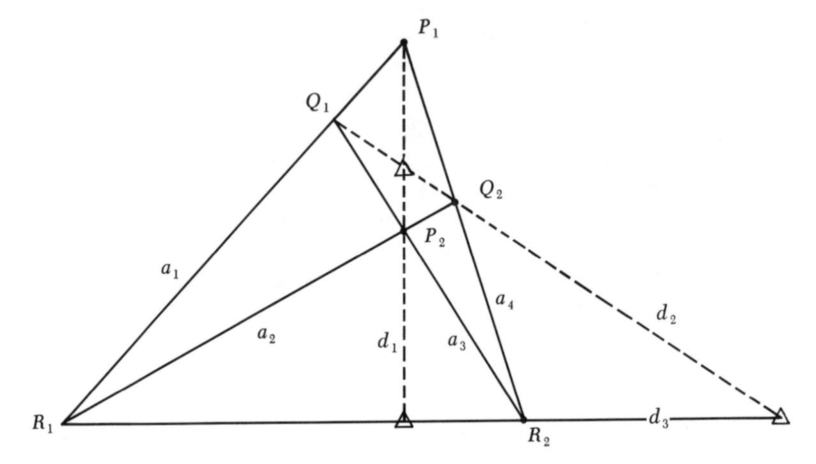

FIG. 5.14 The complete quadrilateral.

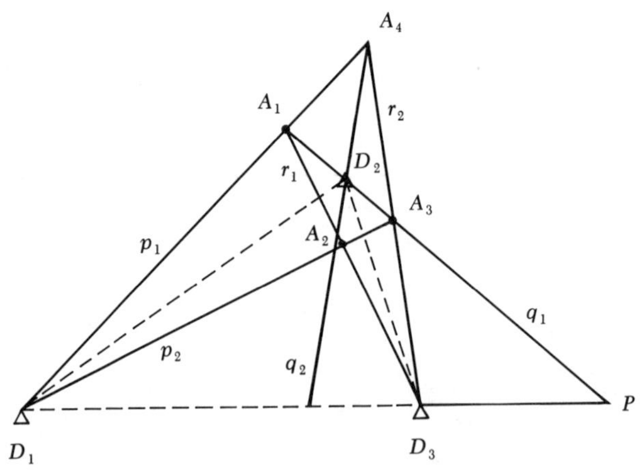

FIG. 5.15 The complete quadrangle.

The special notation given below is used in describing the dual Figs. 5.14 and 5.15.

(A_i,A_j) for the line on points (A_i and A_j)

(a_i,a_j) for the point on lines (a_i and a_j)

$\begin{pmatrix} A_i,A_j \\ A_k,A_m \end{pmatrix}$ for the point on lines (A_i,A_j) and (A_k,A_m)

$\begin{pmatrix} a_i,a_j \\ a_k,a_m \end{pmatrix}$ for the line on points (a_i,a_j) and (a_k,a_m)

The Complete Quadrilateral		***The Complete Quadrangle***	
4 *sides* (lines) a_1, a_2, a_3, a_4		4 *vertices* (points) A_1, A_2, A_3, A_4	
6 *vertices* (points)		6 *sides* (lines)	
Vertex	Opposite vertex	Side	Opposite side
$(a_1a_4) = P_1$	$(a_2a_3) = P_2$	$(A_1A_4) = p_1$	$(A_2A_3) = p_2$
$(a_1a_3) = Q_1$	$(a_2a_4) = Q_2$	$(A_1A_3) = q_1$	$(A_2A_4) = q_2$
$(a_1a_2) = R_1$	$(a_3a_4) = R_2$	$(A_1A_2) = r_1$	$(A_3A_4) = r_2$
3 *diagonal lines*		3 *diagonal points*	
$d_1 = \begin{pmatrix} a_1 & a_4 \\ a_2 & a_3 \end{pmatrix}$		$D_1 = \begin{pmatrix} A_1 & A_4 \\ A_2 & A_3 \end{pmatrix}$	
$d_2 = \begin{pmatrix} a_1 & a_3 \\ a_2 & a_4 \end{pmatrix}$		$D_2 = \begin{pmatrix} A_1 & A_3 \\ A_2 & A_4 \end{pmatrix}$	
$d_3 = \begin{pmatrix} a_1 & a_2 \\ a_3 & a_4 \end{pmatrix}$		$D_3 = \begin{pmatrix} A_1 & A_2 \\ A_3 & A_4 \end{pmatrix}$	

5.7 The quadrangle axiom

In the finite geometry illustrated in Fig. 2.2 the line contained only three points, and the quadrangle with vertices 1, 4, 7, 6 had 2, 3, 5 for its diagonal points. Point 2 was the intersection of lines (1,4) and (7,6); point 3 of lines (1,7) and (4,6); and point 5 of lines (1,6) and (4,7). Figure 2.2 shows these three points lying on a (dotted) line.

The next axiom numbered in sequence with the earlier axioms 1–8 rules out the possibility of diagonal points being collinear.

The Quadrangle Axiom 9 *The diagonal points of a quadrangle are not collinear.*

This new axiom is used now in proving the following theorem:

Theorem 5.7 *On a line are at least four points.*

proof: Let a side A_1A_3 through the diagonal point D_2 of quadrangle $A_1A_2A_3A_4$ (Fig. 5.15) intersect the line D_1D_3 in a point P. Then point P cannot coincide with A_1 or A_3, for if it did, three vertices of the quadrangle would be collinear contrary to hypothesis; and P cannot coincide with D_2 since if it did, the diagonal points D_1, D_2, D_3 would be collinear. Thus the four collinear points A_1, D_2, A_3, P are distinct. They determine with a point S not on their line four distinct lines intersecting any line L' of the plane not through S in four distinct points A_1', D_2', A_3', P and the theorem is proved. Its dual is the theorem:

Theorem 5.8 *On a point are at least four lines.*

5.8 Perspective quadrangles and quadrilaterals

Definitions 1 and 2 of Sec. 5.4, when extended to polygons of $n \geq 3$ sides, provide definitions of perspective polygons. A useful property of perspective quadrangles is established in the next theorem. Its dual supplies the corresponding property for perspective quadrilaterals.

Theorem 5.9 *If two complete quadrangles are so situated that five pairs of corresponding sides intersect in points of one straight line, then the sixth pair of corresponding sides will intersect in a point of the same line.*

Let the quadrangles $A_1A_2A_3A_4$ and $A_1'A_2'A_3'A_4'$ (Fig. 5.16) be so situated that the five points P, Q, R, S, T:

$$P = \begin{pmatrix} A_1A_4 \\ A_1'A_4' \end{pmatrix} \qquad Q = \begin{pmatrix} A_2A_3 \\ A_2'A_3' \end{pmatrix} \qquad R = \begin{pmatrix} A_2A_4 \\ A_2'A_4' \end{pmatrix} \qquad S = \begin{pmatrix} A_3A_4 \\ A_3'A_4' \end{pmatrix}$$

$$T = \begin{matrix} A_1A_2 \\ A_1'A_2' \end{matrix}$$

of intersection of pairs of corresponding sides lie on line L. The theorem will be proved by showing that the point U of intersection of the sixth pair of corresponding sides (A_1A_3), $(A_1'A_3')$ is also on L.

Since the triangles $A_2A_3A_4$ and $A_2'A_3'A_4'$ have line L as an axis of perspec-

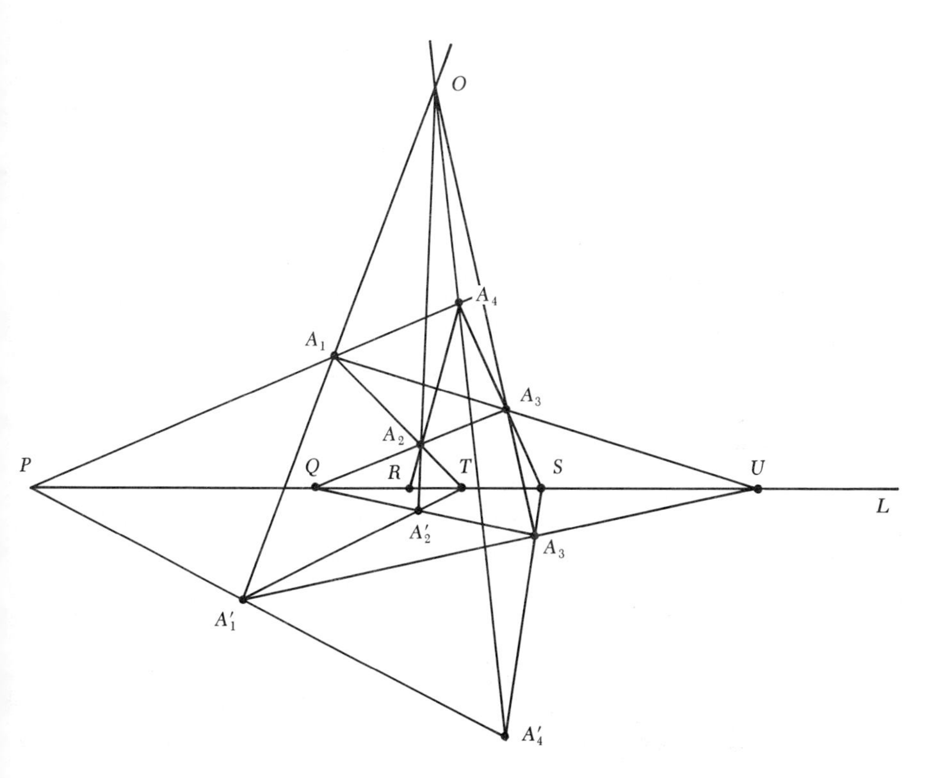

FIG. 5.16

tivity, it follows from Desargues' theorem that the lines A_2A_2, A_3A_3', A_4A_4' joining corresponding vertices are concurrent at a point, say O. But the triangles $A_1A_2A_4$ and $A_1'A_2'A_4'$ also have line L as an axis of perspectivity, and since the lines A_2A_2' and A_4A_4' are already known to meet at O, line A_1A_1' also passes through O. The triangles $A_1A_2A_3$ and $A_1'A_2'A_3'$ are therefore perspective from point O. But line L is the axis of perspectivity of these last two triangles, and hence the corresponding sides A_1A_3 and $A_1'A_3'$ intersect in point U of L. Thus the theorem is proved.

From the principle of duality follows the dual theorem:

Theorem 5.10 *If two complete quadrilaterals are so situated that five of the lines joining pairs of corresponding vertices pass through one point, the lines joining the sixth pair of corresponding vertices will pass through the same point.*

Exercise 5.4

1 State the plane dual of Axiom 9.

2 Does a quadrilateral in the Euclidean plane have six vertices when the quadrilateral is a parallelogram? Why?

3 Given the diagonal triangle and one vertex of a quadrangle, show how to locate with straightedge only the remaining three vertices *Hint:* See ([32] p. 17).

4 Draw a figure illustrating Theorem 5.10 and prove the theorem without using the principle of duality.

5 Show that the diagonal triangle of a quadrangle is perspective with each of the four triangles formed by a set of three of the vertices, the center of perspectivity being in each case the fourth vertex.

Concluding remarks

The entire theory of this chapter has been based on only nine axioms: the eight of Sec. 5.1 and the quadrangle axiom of Sec. 5.7. Through their use a number of significant facts have been established. Proof has been given that a line contains *at least four points,* and Desargues' theorem has been proved without the use of the metric concepts which entered into the earlier proof. The latter proof is indeed a most surprising and unexpected result.

After a statement and illustrations of the principle of duality, the dual figures, quadrangles and quadrilaterals, were described in detail. Then perspective polygons were defined, and attention was called to an important property of perspective quadrangles and quadrilaterals. This property is used in the next chapter, where harmonic elements are introduced in a nonmetric setting.

A moment's reflection shows that more axioms are needed if the line is to contain more than four points. These new axioms will be given later, but attention is called first to the plan of this development. A few axioms are given at the start, and some of their consequences are deduced before others are given. In this way one is made aware of the purpose served by each new set.

Building or creating a new geometry is much like any other construction project. The foundation (axioms) comes first and then the superstructure (theorems). However beautiful the finished product may be, its durability and usefulness will be determined by the strength of its foundation. Sometimes weaknesses are hard to find, as the discussion of Chap. 2 pointed out.

6 projective theory of harmonic elements: additional axioms

Much of interest was learned earlier about harmonic points and lines in the Euclidean plane. Even more can be learned about them in the projective plane. The purely projective definitions of harmonic points and lines given in this chapter are in sharp contrast with the earlier ones (Sec. 4.7) which involved metric concepts. As in the case of Desargues' theorem, these metric concepts are superfluous and tend to disguise the inherent simplicity and beauty of the harmonic concept.

After the new definitions have been given, there will be added to the present axiomatic structure another set of assumptions called separation axioms. The new axioms, which, incidentally, take the place of order axioms in Euclid, will be used to show that a line contains an infinite number of points. Only one final axiom will then be needed to show that points of the projective line may be put into 1-1 correspondence with elements of the real number system, including the symbol ∞.

This is the preparation, among other things, for attaching numbers, called coordinates, to points and lines of the plane.

6.1 Projective definition of harmonic elements

It was shown in Fig. 5.16 that a line not through a vertex of a quadrangle meets the six sides of the quadrangle in the six distinct points P, Q, R, T, S, U, which may or may not all be distinct. Special interest is attached to these points when they are not all distinct and the six points become the four points A, B, C, D. Then the line through two of the diagonal points A and B of the quadrangle $PQRS$ (Fig. 6.1) passes through the points C and D in which the sides through the remaining diagonal point O meet this line. The four points A, B, C, D are then called harmonic, in accordance with the formal definition:

Definition 1 *Four points A, B, C, D on a straight line are harmonic when they are so related that two of them A, B are diagonal points of a quadrangle, while the remaining two C, D are the points in which*

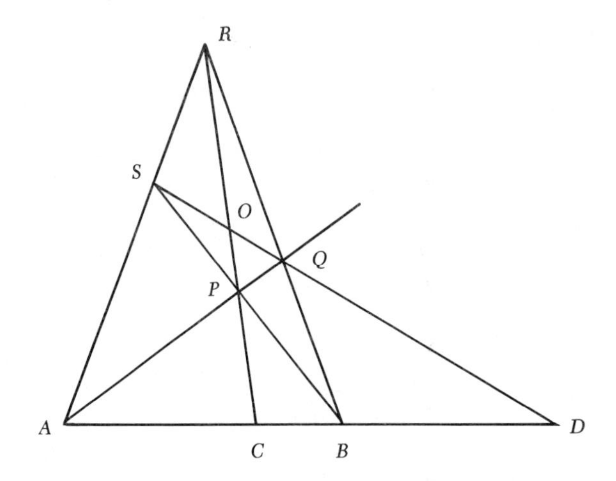

FIG. 6.1 Harmonic points $H(AB,CD)$.

the sides passing through the third diagonal point meet the line AB (Fig. 6.1).

Dual to the notion of a harmonic set of points is a harmonic set of lines.

Definition 2 *Four lines a, b, c, d on a point are harmonic when they are so related that two of them c, d are diagonals of a quadrilateral, while the remaining two a, b are lines joining the point H of meeting of these two diagonals to the vertices on the third diagonal (Fig. 6.2).*

FIG. 6.2 Harmonic lines $H(ab,cd)$.

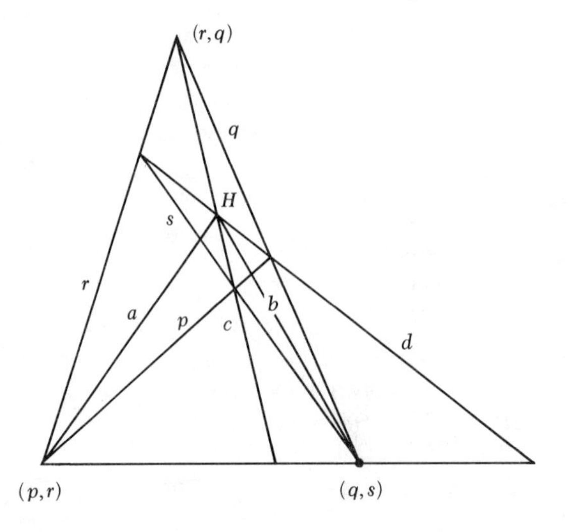

6.2 Harmonic conjugates

The projective definition of harmonic points given in Sec. 6.1 is a far cry from the earlier one of Sec. 4.7 which used metric concepts. But, certain similarities appear on closer observation. The new, as well as the old, definition automatically separates the four points A, B, C, D (Fig. 6.1) into the pairs A, B and C, D. Such a pairing is denoted by the symbol $H(AB,CD)$.

Either member of the pair (C,D) is called the harmonic conjugate of the other member with respect to the remaining pair (A,B). Thus C (D) is the harmonic conjugate of D (C) with respect to the pair (A,B).

Dually, $H(ab,cd)$ denotes the pairing of the harmonic lines (Fig 6.2) into the two pairs (a,b), (c,d). Again, either *member of the pair (c,d) is the harmonic conjugate of the other member with respect to the remaining pair (a,b).*

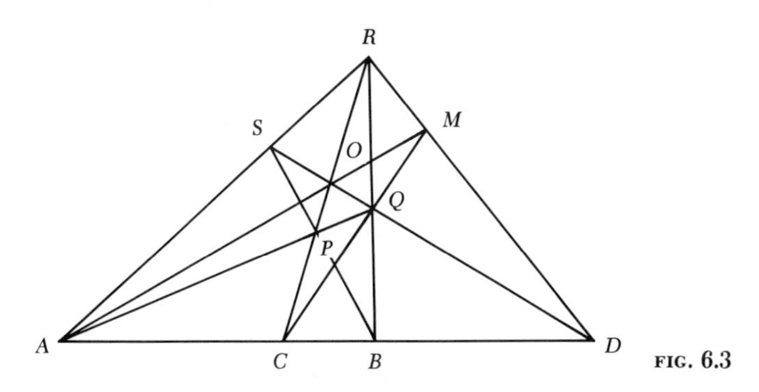

FIG. 6.3

It makes no difference which pair in a harmonic set of elements is called the harmonic conjugate of the other pair. To prove this statement, it is noted first that in the harmonic set $H(AB,CD)$ of Fig. 6.3, A, B, O are diagonal points of the quadrangle $PQRS$. If lines CQ and DR intersect at M, the three points

$$D = \binom{SO}{RM} \qquad C = \binom{PO}{QM} \qquad B = \binom{SP}{RQ}$$

are collinear, and the triangles POS and RQM are perspective from A. By Desargues' theorem, the line OM passes through A. Therefore, in the quadrangle $OQMR$, C and D are diagonal points, while sides through its third diagonal point meet CD in points A and B. This means that A, B are harmonic conjugates with respect to points C and D, and proof has been given of the theorem:

Theorem 6.1 *If points C, D are harmonic conjugates with respect to points A, B, then points A, B are harmonic conjugates with respect to points C, D.*

From the principle of duality follows the dual theorem:

Theorem 6.2 *If lines a, b are harmonic conjugates with respect to lines c, d, then lines c, d are harmonic conjugates with respect to lines a, b.*

From the definitions just given and Theorems 5.9 and 5.10 follows the next theorem:

Theorem 6.3 *The harmonic conjugate of a point (line) with respect to two other points (lines) is uniquely determined.*

6.3 The invariance of the harmonic property

As stated earlier, the great importance of the harmonic property lies in the fact that it is invariant on projection. This is essentially what the next theorem says.

Theorem 6.4 *The lines joining a harmonic set of points to a point not on the line of the points is a harmonic set of lines; and conversely the points of meeting of a harmonic set of lines with a line not through their point of meeting is a harmonic set of points.*

proof: Let points A, B, C, D (Fig. 6.4) form a harmonic set $H(AB,CD)$, and let S be any point not on their line. Through a point A draw a line meeting SC and SB in points P and Q respectively, and let line BP meet SA in R. Then, the quadrangle $SRPQ$ has diagonal points at A and B, and one of its sides passes through C. By Theorem 5.9 the sixth side RQ passes through the point D; and the lines AD, AQ, RD, RB form a quadrilateral of which SA and SB are diagonals.

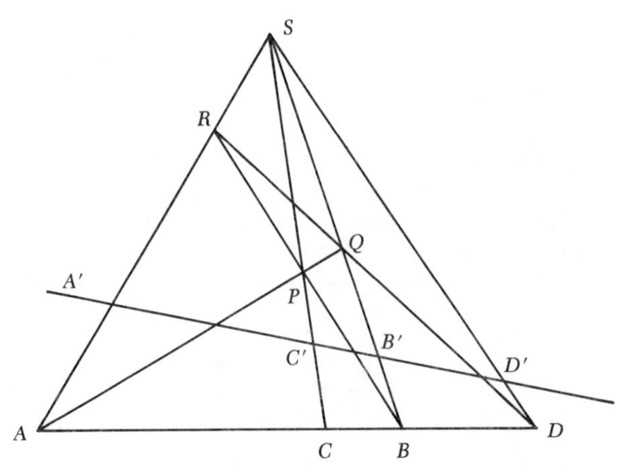

FIG. 6.4

Since SC and SD are lines from S to the vertices P, D lying on the third diagonal, the lines SA, SB, SC, SD are harmonic.

Conversely, if a transversal intersects harmonic lines through S in points A, B, C, D, these points form the harmonic set $H(AB,CD)$, because, if a line through A meets SC and SD in points P, Q, the quadrilateral with sides AD, DQ, PQ, PB has SA and SB as two of its diagonals. The sides DQ and PB intersect, therefore, in a point R of SA; and in the quadrangle $PQRS$, A and B are diagonal points. Sides through the third diagonal point therefore meet line AB in points C and D, and by definition 1 of Sec. 6.1, the four points A, B, C, D form the harmonic set $H(AB,CD)$. Thus the theorem is proved.

6.4 Straightedge constructions

In elementary geometry, a construction problem involves an explanation of how the straightedge and compass[1] may be used to obtain new points and lines, which may in turn determine other desired points and lines.

The classical restrictions to these two instruments were apparently suggested by the following Euclidean axioms:

1 A straight line may be drawn from any point to any point.
2 A finite straight line may be produced continuously in a straight line.
3 A circle may be drawn with any center and any radius.

Constructions made in accordance with these axioms are called ruler-and-compass constructions. By the ruler is here meant an unmarked straightedge. The modern ruler is a graded straightedge.

As Euclid's axioms have been interpreted, they place restrictions not only on the instruments to be used in the construction but also on the manner in which they are to be used. The straightedge is to be used for drawing a line through two given points and extending it in either direction. That is all.

The third axiom tells how the compass is to be used. When Euclid drew a circle, probably on the sand, the center and a point on the circumference had to be given. Then it was as if the compass automatically closed, so that this position of the compass could not be used to draw another circle. However, Euclid's second axiom shows how to construct at any given point a line segment equal to a given line segment. Circles were then drawn by using a radius equal to a given line segment.

With the straightedge and compass, the ancient Greeks were able to solve a variety of problems. They bisected lines and angles, drew perpendiculars to lines, erected tangents to circles, and constructed certain regular poly-

[1] The words "compass" and "compasses" are here taken to be synonymous.

gons. They even disposed of the famous problem of Apollonius (about 250 B.C.), in which three arbitrary circles in the plane are given and a fourth circle tangent to all three is required.

In the present treatment of projective geometry, in which distance has not been defined, the compass can have no part in construction problems, but it is interesting to observe that many straightedge- (ruler-) compass constructions of elementary geometry may be replaced by constructions using only a straightedge. One such construction is given for the following problem:

Construction Problem When three points A, B, C of a harmonic set of points are given, construct the fourth point D, the harmonic conjugate of C with respect to A and B.

To perform the construction, draw through any one of the given points, say A (Fig. 6.5), a line L' distinct from the line L of the given points; and on L' select two points U, V distinct from A, and join U to C and V to B. Then if lines CU and BV meet at R_1 and lines AR_1 and BU at R_2, in symbols,

$$R_1 = \begin{pmatrix} BV \\ CU \end{pmatrix} \qquad R_2 = \begin{pmatrix} AR_1 \\ BU \end{pmatrix}$$

the intersection of lines L and VR_2 is the required point D:

$$D = \begin{pmatrix} AB \\ VR_2 \end{pmatrix}$$

This is so because A and B are diagonal points of the quadrangle R_1R_2UV

FIG. 6.5

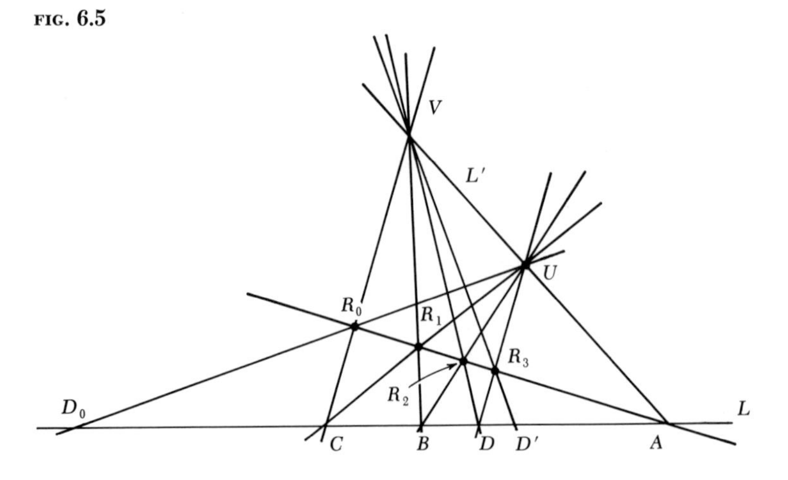

(Fig. 6.5), and sides VR_2 and UR_1 through the remaining diagonal points intersect L in points C, and D conjugates with respect to A and B.

If now

$$R_3 = \binom{AR_2}{DU}$$

the intersection of lines L and VR_3 is a point D' on L such that $H(AD,D'B)$, where D' is distinct from A and B. But is D' distinct from point C? If it is, the construction may be continued to locate other points of the line. Obviously, the construction cannot continue to yield new points indefinitely if the number of points on the line is finite, for eventually a constructed point must coincide with one of the given points or a previously constructed point.

Exercise 6.1

1. Show that the position of point D (Fig. 6.5) is independent of the choice of points U and V on L'.
2. For what position of the points A, B, C of Fig. 6.5 will the construction given in Sec. 6.4 fail when L is a line of the Euclidean plane? Explain.
3. If, in Fig. 6.5, lines VC and AR_1 intersect at R_0 and lines R_0U and L at D_0, is $H(AC,BD_0)$? Why?

6.5 Separation axioms

So far in the axiomatic development of projective geometry it is known only that a line contains at *least four points*. If there are to be others, as is desirable, their existence must be postulated by additional axioms, and the new axioms must be consistent with facts already established.

A set of new axioms which guarantees the existence of an infinite number of points on the line is linked with the common notion of "separation" illustrated many times in everyday life in such cases as one room of a house being separated from another by a wall, two countries being separated by a mountain range, a piece of string being separated into two parts by a single cut, or a circular wire being separated into two parts by two cuts.

On a Euclidean line, a single point C separates a point A from a point B (Fig. 6.6*a*) in the sense that starting at either point and moving continuously in the same direction one encounters the point C. This situation is quite different, however, if the three points are situated on a circle (Fig. 6.6*b*), where the points A and B are extremities of two different segments: one, the heavy line segment, and the other, the dotted line segment. If C is a point on one of them, say the heavy line segment, one can move from A to B on the dotted segment without encountering C. But, if there is a point D

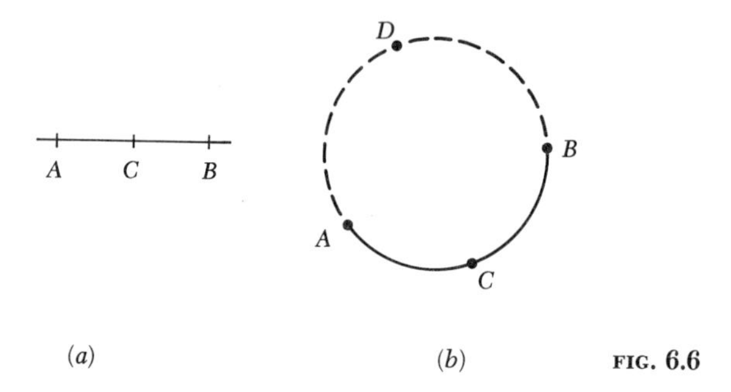

FIG. 6.6

located on this same dotted segment, one cannot move from A to B in *either direction* without encountering one or the other of the points C or D. That is why *two points* rather than a single one are needed to separate two points on a circle from another pair of points on the circle.

Again, suppose that O is a point on a line L in the Euclidean plane, and P the center of a circle tangent to L at O (Fig. 6.7). Then any line through O forming an angle $\theta \neq 90°$ with PO meets L in a point A. The line making angle $\theta = 90°$ with PO is parallel to L and hence does not meet it. This lack of a 1-1 correspondence between lines through P and points of line L is like a nonexistent bridge from one side of a river to the other. There is no way by which a point A moving on the Euclidean line L to the right of a point O can reach a point to the left of O.

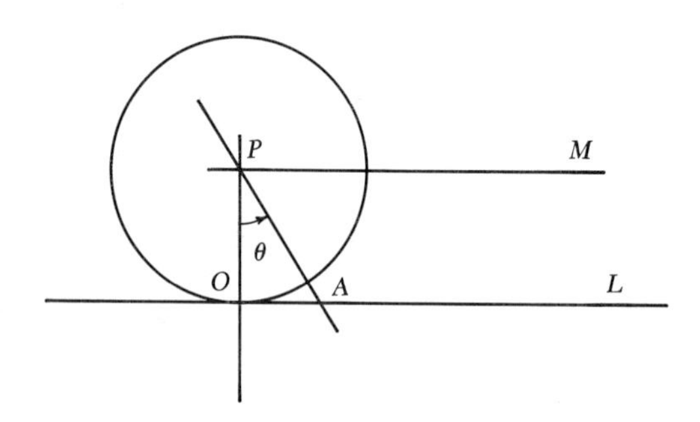

FIG. 6.7

In the projective plane, however, the situation is quite different. There is a bridge to the other side, for every line through a point P of the projective plane meets line L in a point A, and A may move continuously from a point O, and continuing in the same direction, A may eventually return to the starting point O just as if it were moving on a circle.

What this all comes down to are heuristic *aids* to an understanding of the following axioms which deal with an undefined relation called *separation.* Since they are additions to the present axiomatic structure, the new axioms are numbered in sequence with the earlier ones of Secs. 5.1 and 5.7.

Separation Axioms 10 *The pairs A, B and C, D of a harmonic set of points H(AB,CD) separate each other.*

11 *If the pairs A, B and D_1, C separate each other and if also the pairs A, D_1 and D_2, B separate each other, then the pairs A, B and C, D_2 separate each other.*

12 *If the pairs A, B and C, D separate each other, then A, B, C, D are distinct points.*

The axioms are obviously describing an arrangement of points on a circle (Fig. 6.8).

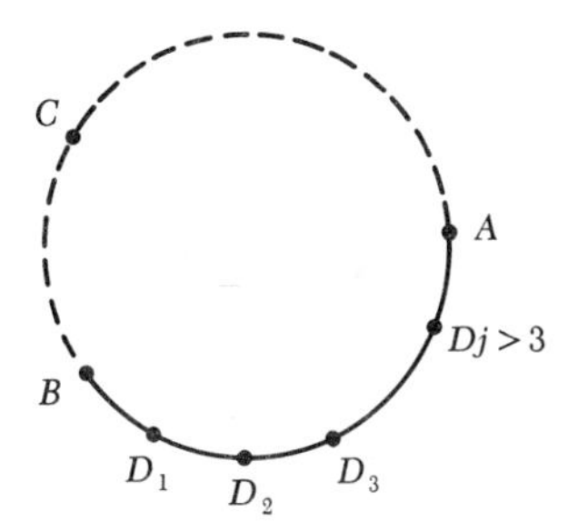

FIG. 6.8

The symbol $AB \,//\, CD$ is used to denote the separation of the pair of points A, B by the pair C, D, and a segment of the projective line may now be defined without the use of metric concepts: *If A, B, C are three fixed points of the line L, the set of all points X such that AB // CX is a segment of the line having A, B as its extremities.*

The symbol $AB \,/\, C$ is used to denote a segment not containing the point C.

The new terminology and symbolism are used in the following construction problem.

6.6 Another construction problem

Construction Problem If points A, B are extremities of a given segment of the projective line, locate, using straightedge only, a point D on the other segment of the line.

To perform the construction, choose an arbitrary point C of the given line segment (Fig. 6.9), and construct two quadrangles $P_1Q_1R_1S_1$ and $P_2Q_2R_2S_2$, one above and the other below the segment. Then, from definition 1 of Sec. 6.1 and Theorem 6.3, each of the lines S_1Q_1 and S_2Q_2 intersects line AB in the point D such that

$$H(AB,CD)$$

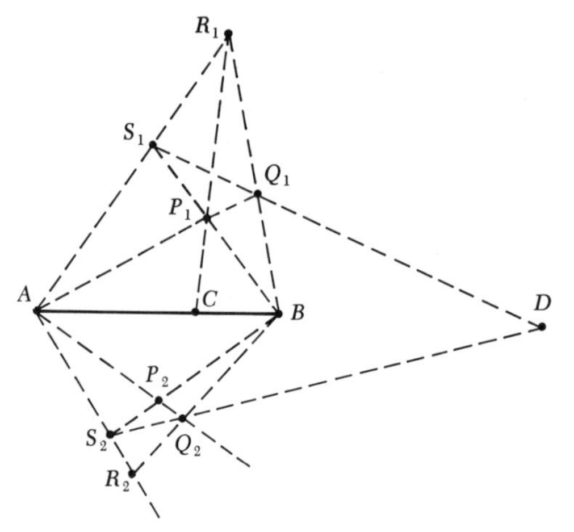

FIG. 6.9

But then, it follows from Axiom 10 that

$$AB \,//\, CD$$

and hence D is on the line segment $AB\ /\ C$ as was required.

6.7 A harmonic sequence

The separation axioms will be used now to show that the projective line contains an infinite number of points

$$D_1, D_2, D_3, \ldots, D_n$$

satisfying the condition

$$H_1(AB,D_1C), H_2(AD_1,D_2B), \ldots, H_n(AD_{n-1},D_{n-2}D_n)$$

and forming a harmonic sequence in accordance with the definition:

> *A set of points $D_1, D_2, \ldots, D_n$ on a line L through the three fixed points A, B, C forms a harmonic sequence if point D_1 is the harmonic conjugate of one of the points A, B, C with respect to the other two, and any following point D_j is the harmonic conjugate of any one of the preceding points of the sequence with respect to two other preceding points of the sequence.*

The separation axioms are used now to show that each point D_j of the harmonic sequence $A, B, C, D_1, D_2, \ldots, D_n$ is distinct from its predecessor.

To do so, it will suffice to show that for any integer $1 < j \leq n$, D_j is distinct from D_r, $1 \leq r < j$.

Consider the k-relations:

$$AD_j \,//\, D_{j+1}D_{j-1} \tag{1}$$
$$AD_{j-1} \,//\, D_jD_{j-2} \tag{2}$$
$$AD_{j-2} \,//\, D_{j-1}D_{j-3} \tag{3}$$
$$\cdots\cdots\cdots\cdots\cdots\cdots$$
$$AD_{j-k+1} \,//\, D_{j-k+2}D_{j-k} \tag{k}$$

From (1) and Axiom 12, *D_j is distinct from D_{j-1}*. From (1) and (2) and Axiom 11,

$$AD_j \,//\, D_{j+1}D_{j-2} \tag{k + 1}$$

and D_j is *distinct also from D_{j-2}*.

From (2) and (3) and Axiom 11,

$$AD_{j-1} \,//\, D_jD_{j-3} \tag{k + 2}$$

which, with (1) and Axiom 11, gives

$$AD_j \,//\, D_{j+1}D_{j-3} \tag{k + 3}$$

and D_j is *distinct also from D_{j-3}*.

Similarly, it is shown through the use of the $k = j - 1$ relations and Axiom 11,

$$AD_j \,//\, D_{j+1}D_1$$

and D_j is *distinct also from D_1*. Since the points $D_1, D_2, \ldots, D_j$ are infinite in number, the line L contains an infinite number of points; and since a 1-1 correspondence exists between line L and any other line of the plane, proof has been given of the theorem:

Theorem 6.5 *Every line of the projective plane contains an infinite number of points.*

Exercise 6.2

1 Select three fixed points A, B, C on a line and construct the point D_1 such that $H(AB,CD_1)$ and the point D_2 such that $H(AD_1,BD_2)$. Is D_2 distinct from D_1? Prove your answer.

2 If D_3 is a third point of the line of question 1 such that $H(AD_2,D_1D_3)$, is D_2 the midpoint of the line segment D_1D_3? Can a point C be selected on the line AB such that D_2 is the midpoint? Why?

6.8 A continuity axiom

One of the most interesting developments in projective geometry is the attaching of a number called its coordinate to each point of a line without using the distance concept.

It is not the intent to go into this development here since to do so would require too great a digression, and besides, coordinate projective geometry is to be introduced later in a way more in keeping with the plan of the text. However, a few remarks in this connection will motivate the addition of the final axiom to the present set of axioms.

Proof has just been given that on the projective line L through the three fixed points A, B, C are points $D_1, D_2, \ldots, D_n$ of a harmonic sequence determined by the points. A 1-1 correspondence will be set up between these points and the set of integers $-n, \ldots, -1, 0, 1, \ldots, n$.

First, the coordinates 0, 1, ∞ are *arbitrarily* attached to the respective points A, B, C, and then to the successive points $D_1, D_2, \ldots, D_n$ (Fig. 6.10) are attached the respective coordinates $2, 3, \ldots, n$. This means that if the point corresponding to the integer n be denoted $P(n)$, $P(1)$ is the harmonic conjugate of $P(\infty)$ with respect to $P(0)$ and $P(2)$; $P(3)$ is the harmonic conjugate of $P(\infty)$ with respect to $P(2)$ and $P(4)$, etc. Also, $P(0)$ is the harmonic conjugate of $P(\infty)$ with respect to $P(-1)$ and $P(1)$; $P(-1)$ is the harmonic conjugate of $P(\infty)$ with respect to $P(-2)$ and $P(0)$, etc.

If n be called the coordinate of $P(n)$, a glance at Fig. 6.10 shows that points $P(0)$, $P(1)$, $P(2)$, $P(3)$, . . . are not evenly spaced as they are when the

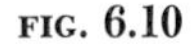
FIG. 6.10

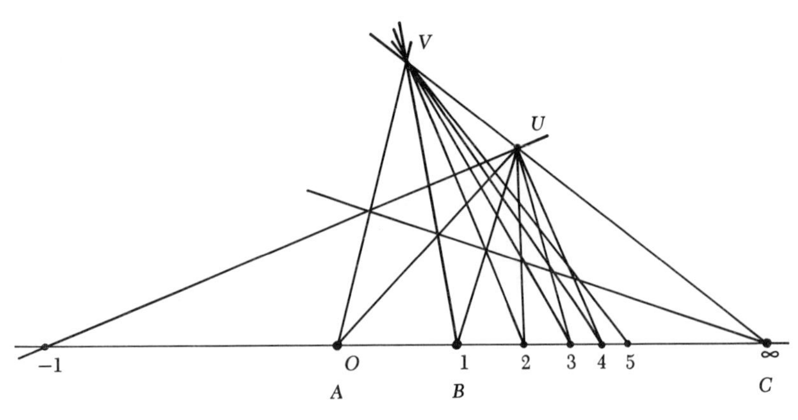

coordinate represents the distance from a fixed point of the line. It is instructive to see what happens to the spacing as the point C is taken further and further out on the line (see Exercise 6.2, question 2). Moreover, there are gaps between the points pictured. If there are points between them, as is desired for future theory, their existence must be postulated. This has been done. By suitably defining the sum and product of points ([100] Chap. 7) the gaps are filled in and to each rational number m/n, with m, n positive integers, corresponds a point $P(m/n)$ of the line. The resulting set of points is called the *net of rationality determined by the three fixed points* $P(0)$, $P(1)$, $P(\infty)$.

Although the new set is everywhere dense, the process has not provided for points $P(x)$, where x is an irrational number, and they are needed if full advantage is to be taken of existing algebraic theory.

One final axiom—a continuity axiom—will take care of this matter.

A Continuity Axiom 13 *There exists a projective line L containing a set of points isomorphic with the set of numbers of the extended real number system.*

Because of this axiom it is possible to set up a 1-1 correspondence between points of the projective line and numbers of a number system of which the most familiar example is the real number system.

A number x associated with a point P of the line is called its coordinate, and the point is then indicated $P(x)$.

When the number system employed is real, the point $P(x)$ is said to be real, and the point associated with the symbol ∞ is usually called the ideal point $P(\infty)$ of the line.

Exercise 6.3

1 Prove the theorem: "If a projective geometry has at least one line of rational points, every line is a line of rational points."

2 Consider the three fixed collinear points $A(0)$, $B(1)$, $C(3)$, where coordinates of points are distances from a fixed point of the line. Construct, using straightedge only, the point D such that $H(AB,CD)$ and the point D' such that $H(AC,BD')$, and give coordinates of each.

3 Given three rays of a pencil, find the fourth ray which is the harmonic conjugate of a particular one of the three, relative to the other two.

4 Using the straightedge only, construct through a given point in the projective plane a line which will pass through the inaccessible point of intersection of two given lines in the plane.

5 A line intersects the sides of a triangle ABC in the points A_1, B_1, C_1, and the harmonic conjugates A_2, B_2, C_2 of these points are determined relative to the

two vertices on the same side. Show that the points A_1, B_2, C_2 are collinear as are also the points B_1, A_2, C_2' and the points C_1, A_2, B_2. Name triples of concurrent lines.

Concluding remarks

It has taken 13 axioms in all to reach the place where points of a line may be put into 1-1 correspondence with numbers of the extended real number system and algebraic processes made available for the geometer. They are:

Axioms 1–8: Incidence and Existence Axioms
Axiom 9: The Quadrangle Axiom
Axioms 10–12: Separation Axioms
Axiom 13: A Continuity Axiom

Harmonic elements which figured prominently in the development were used in the statement of the separation axioms. These axioms, which replaced order axioms in Euclid, were then used to show that a line contained an infinite number of points. A continuity axiom then permitted setting up a 1-1 correspondence between points of the line and the extended real number system.

One final chapter on synthetic pure geometry will round out the announced intention of showing what can be accomplished by purely synthetic, nonmetric methods. The results are impressive. A truly magnificent structure has been built without benefit of the mortar and steel supplied by metric notions.

Suggestions for further reading

Fishbach, W. T.: "Projective and Euclidean Geometry," chap. 10, Axiomatic Projective Geometry.
Holgate, T. F.: "Projective Geometry," chap. 3, Harmonic Forms.
Meserve, B. E.: "Fundamental Concepts of Geometry," chap. 3, Separation Axioms. Nets of Rationality. Real Projective Geometry.
Robinson, G. de B.: "The Foundations of Geometry," chap. 7, Addition and Multiplication of Points on a Line. Order and Continuity.
Seidenberg, A.: "Lectures in Projective Geometry," chap. 3, Geometric Description of the Operations Plus and Times.
Young, J. W.: "Projective Geometry," chap. 8, The Algebra of Points and the Introduction of Analytic Methods.

7 perspectivities, projectivities, and the projective theory of conics

What has been selected for presentation here is of special interest to the artist with its emphasis on perspectivities, projectivities, and nonmetric properties of figures. Here, perhaps more than anywhere else in the text, one sees clearly the interplay of mathematics and art. The aesthetic appeal of the material cannot be challenged, nor can its mathematical importance.

This is the geometric background that illuminates and clarifies much of the algebraic theory of later chapters.

7.1 Mappings, transformations, and the function concept

When some rule is given that assigns to each element a of a set A an element b of a set B, A is said to be mapped *into* the set B and b is called the image of a under the mapping.

If every element of B is the image of an element of A, set A is said to be mapped *onto* set B.

The set of ordered pairs (a,b), $a \in A$, $b \in B$, under a mapping from set A to set B, represents a *function* f if no two pairs of the set have the same first element. The set A is then called the *domain of* the function and the totality of image points $b = f(a)$ is its *range*.

A mapping that assigns to each element a of set A a unique element b of set B is said to be *single-valued.*

A telephone directory is a mapping T from a set A of names *into* a set B of numbers (Fig. 7.1). If any name has more than one number attached to

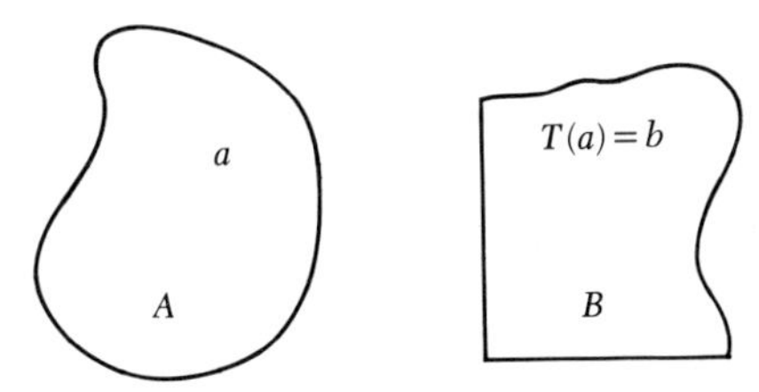

FIG. 7.1

it, in other words, if a person has more than one telephone, the mapping is not single-valued.

A mapping of numbers into numbers is usually shown by means of an equation such as

$$y = f(x) = x^2 \tag{1}$$

where x is a real number and $y = f(x)$ is the value assigned to x by (1). Here the domain of the function is the set A of real numbers, and the range is the set B of positive numbers and zero. The function, or mapping, is single-valued since to each value of x corresponds one and only one value of y, as shown in the graph of the function (Fig. 7.2).

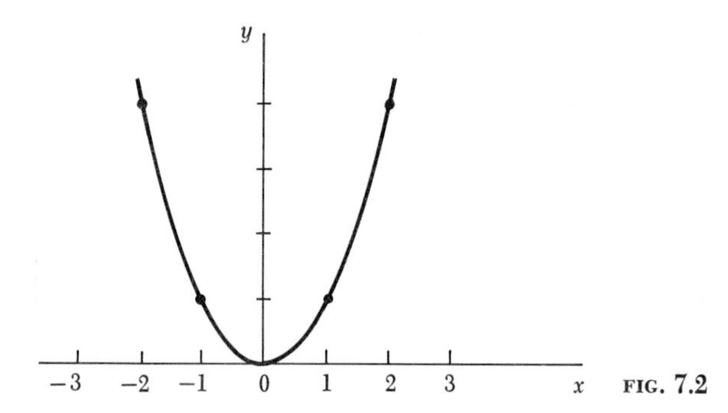

FIG. 7.2

A mapping which is not single-valued is given by the equation

$$y = \pm\sqrt{1 - x^2} \qquad -1 \leq x \leq 1 \tag{2}$$

Here the *domain* is the set A of x values $-1 \leq x \leq 1$, and the *range* is the set B of y values $-1 \leq y \leq 1$, as shown in the graph (Fig. 7.3). To each

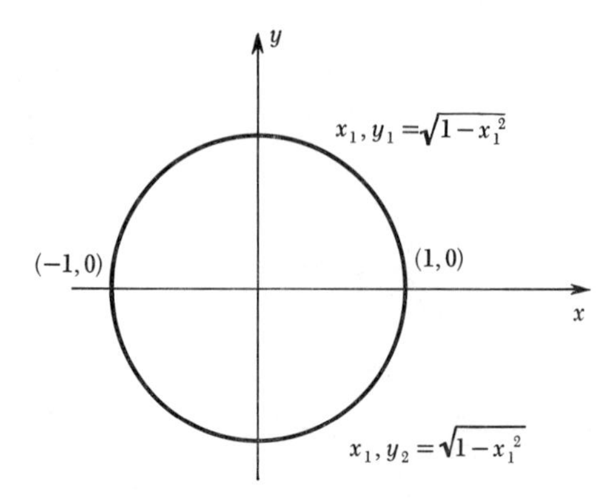

FIG. 7.3

value of x, in the domain, corresponds the two values

$$y_1 = \sqrt{1 - x^2} \qquad y_2 = -\sqrt{1 - x^2}$$

Mappings of interest here are the perspectivities and projectivities discussed next.

7.2 Perspectivities

As is customary in the literature, points of a line will be called a *pencil of points,* and dually, coplanar lines on a point will be called a *pencil of lines.* Since each of these pencils consists of a single infinity of elements, they are known as *one-dimensional forms.*

By definition, a 1-1 *correspondence between two pencils of points is a perspectivity if the lines joining pairs of corresponding points are concurrent at a point S. The point S is called the center of the perspectivity.*

The symbol $\stackrel{S}{\overline{\wedge}}$ is used for the perspectivity with center S. Thus

$$A, B, C, \ldots \stackrel{S}{\overline{\wedge}} A', B', C', \ldots$$

means that points $A, B, C, \ldots$ of line L (Fig. 7.4) are perspective with points $A', B', C', \ldots$ of a second line L' in such a way that A corresponds to A', B to B', C to C', . . . and that the lines AA', BB', CC', . . . are concurrent at S. The point P of meeting of lines L and L' is a self-corresponding point, in that point P of L corresponds to point P of L', and vice versa.

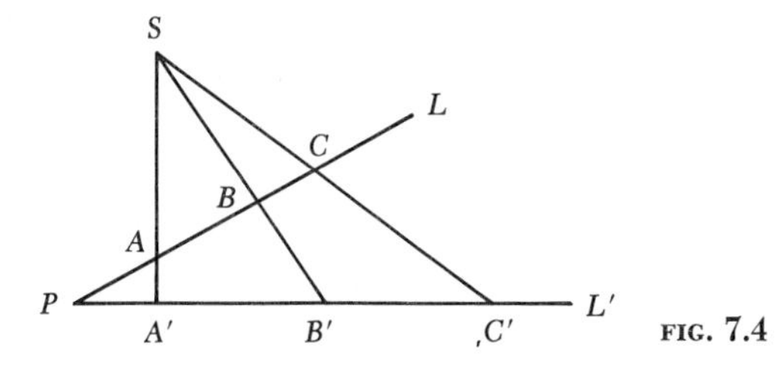

FIG. 7.4

Dually, a correspondence between two pencils of lines is, by definition, a *perspectivity if the points of intersection of pairs of corresponding lines are on a line s, called the axis of the perspectivity.* The symbolic representation

$$a, b, c, \ldots \stackrel{s}{\overline{\wedge}} a', b', c', \ldots$$

means that the pencil of rays $a, b, c, \ldots$ through points S (Fig. 7.5) is perspective with the pencil of rays $a', b', c', \ldots$ through point S' so that the points of meeting of corresponding rays are all on the same line s. The

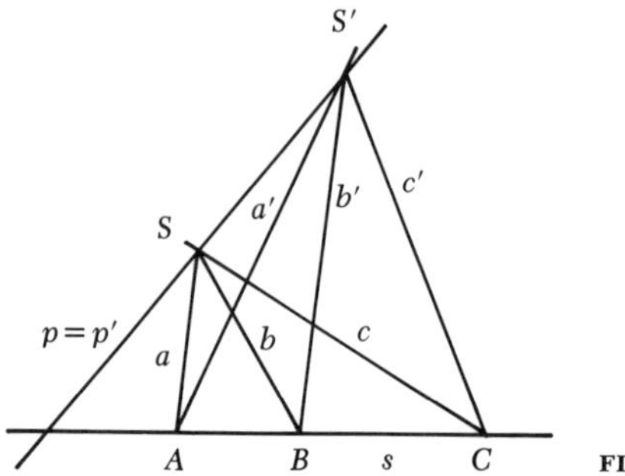

FIG. 7.5

common ray $p = p' = SS'$ is self-corresponding in that line SS' may be considered as belonging to either pencil.

Exercise 7.1

Given a triangle ABC and two distinct points A', B', determine a point C' such that the lines AA', BB', CC' are concurrent at point O and the lines AB', BC', CA' are concurrent at point $O' \neq O$. (The two triangles are then said to be doubly perspective.)

7.3 Projectivities

A discussion of projectivities begins with the formal definition:

> *A correspondence between elements of two one-dimensional forms is a projectivity if such a correspondence is made by a finite sequence of perspectivities.*

A projectivity is illustrated in Fig. 7.6, where points $A, B, C, \ldots$ of line L are perspective through a point S with points $A_1, B_1, C_1, \ldots$ of a second line L_1, and the points of L_1 are in turn perspective through center S_1 with the

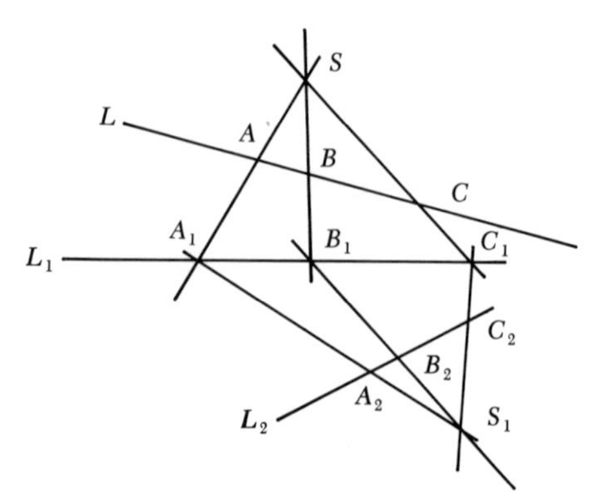

FIG. 7.6

points A_2, B_2, C_2, . . . of a third line L_2. Then, every point of L corresponds to a uniquely determined point of L_2, A to A_2, B to B_2, C to C_2, ... The symbol $\barwedge$ is used to denote a projectivity. Thus

$$A, B, C, \ldots \barwedge A_2, B_2, C_2, \ldots$$

means that points A, B, C, . . . are *projective* with points A_2, B_2, C_2, . . . in such a way that A corresponds to A_2, B to B_2, C to C_2, ... Dually,

$$a, b, c, \ldots \barwedge a_2, b_2, c_2, \ldots$$

means that lines a, b, c, . . . and a_2, b_2, c_2, . . . are projective in such a way that a corresponds to a_2, b to b_2, c to c_2, . . . (Fig. 7.7).

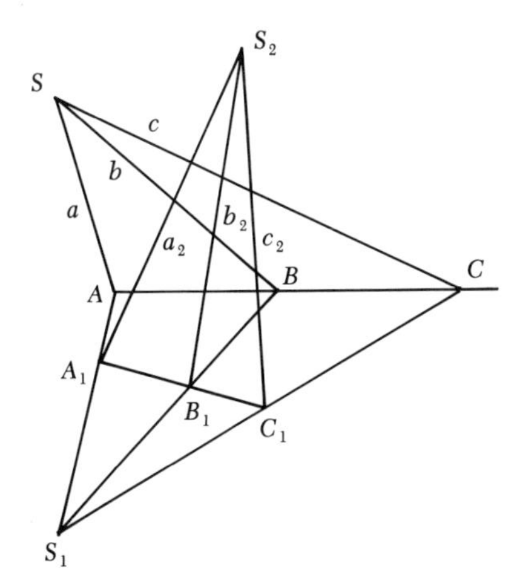

FIG. 7.7

7.4 Theorems on projectivities

It is proposed now to state and prove a set of theorems leading to the fundamental theorem of projective geometry. Interest in these theorems is heightened by the preliminary question:

Question 1 How many sets of corresponding points determine a projectivity?

The question needs some clarification. In a perspectivity, any two points of a line may be made to correspond to any two points of a second line by a suitable choice of the center of perspectivity. To project, for instance, two distinct points A, B of line L (Fig. 7.8) into any two distinct points A', B' of line $L' \neq L$, the center S of perspectivity must be the point of meeting of lines AA' and BB'. After that, a third point C of L *must* correspond to the

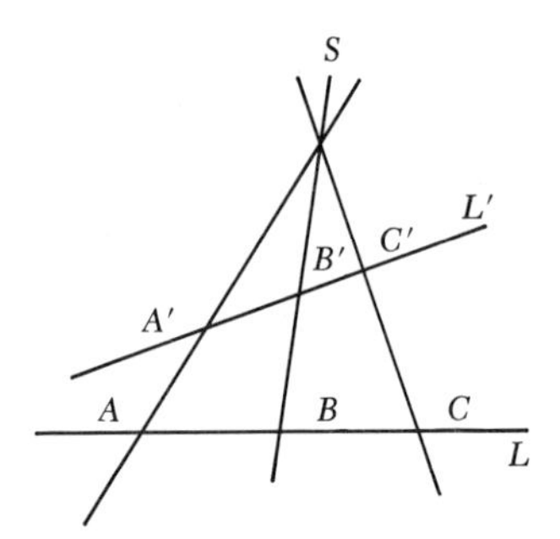

FIG. 7.8

particular point C' in which SC meets line L'. In other words, a third point C of L cannot be sent into an arbitrary point C' of L'. Thus *two pairs of points determine a perspectivity.*

If the projectivity is *not* a perspectivity, the answer to question 1 is by no means obvious. However, there is an initial theorem which shows how any three points of a line may be projected into any three points of another line. A statement and proof of the theorem follows:

Theorem 7.1 *Any three points A, B, C of a line L may be projected into any three points A', B', C' of a line $L' \neq L$ by two centers of perspectivity.*

proof: An auxiliary line L'' is drawn between a pair of noncorresponding points, say A on L and C' on L' (Fig. 7.9), and a point $S_1 \neq C, C'$

FIG. 7.9

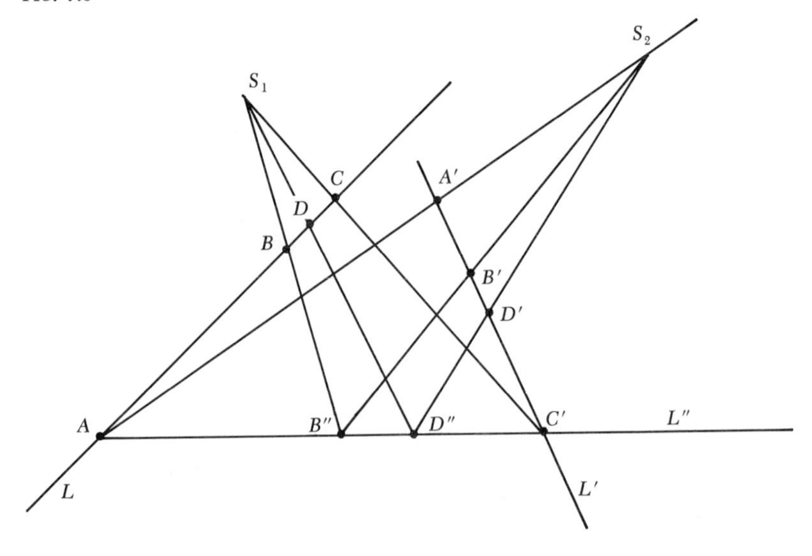

is chosen for the center of a perspectivity from L to L''. If B'' is the projection of point B from S_1, and if the lines AA' and $B'B''$ meet a point S_2, the two points S_1 and S_2 suffice for the desired projectivity. For

$$A, B, C \overset{S_1}{\overline{\wedge}} AB''C' \overset{S_2}{\overline{\wedge}} A'B'C'$$

from which it follows that

$$A, B, C \;\overline{\wedge}\; A'B'C'$$

and the theorem has been proved. The proof is valid for lines which may or may not lie in the same plane.

Corollary *Three centers of perspectivity suffice to project three points of a line into three points of the same line.*

If points A, B, C and their corresponding points A', B', C' under a projectivity are on the *same line* L, the original points A, B, C may be projected into points A', B', C' of a line L'' distinct from L', and the argument just given may be applied to lines L and L' to give the desired projectivity.

From the principle of duality follows the dual theorem and corollary:

Theorem 7.2 *Any three rays a, b, c of a pencil at S may be projected into any three rays of a pencil at S′ by two axes of perspectivity.*

Corollary *Three axes of perspectivity suffice to project three rays of a pencil into three rays of the same pencil.*

That Theorems 7.1 and 7.2 will not hold for *four* elements, instead of *three*, follows immediately from the invariance of the harmonic property on projection.

It is noted also that under the perspectivities from S_1 and S_2 (Fig. 7.9) a point D of line L is projected into a point D' of L'; but because of the arbitrary choice of the auxiliary line L'', there is no guarantee that a different choice of L'' would send point D into the same point D' of L'. This means that question 1 is still unanswered, and the investigations continue.

It may happen that a projectivity

$$A, B, C, \ldots \;\overline{\wedge}\; A', B', C', \ldots$$

established by two or more centers of perspectivity is such that lines AA', BB', CC', . . . joining corresponding points are concurrent and the projectivity is in reality a disguised perspectivity. In more formal language, the projectivity reduces to a perspectivity. The next theorem, which shows when this situation can occur, is an interesting application of Desargues' theorem.

Theorem 7.3 *If three lines L_1, L_2, L_3 are concurrent, the projective correspondence established between points of L_1 and L_3 by a sequence of two perspectivities, one from L_1 to L_2 and the other from L_2 to L_3, may be established by a single perspectivity.*

proof: Let the three given lines L_1, L_2, L_3 meet at the point S (Fig. 7.10). Also, let S_1 be the center of a perspectivity sending points A_1 and B_1 of line L_1 into the respective points A_2, B_2 of line L_2, and let S_2 be the center of a perspectivity sending points A_2, B_2 into points A_3, B_3 of line L_3. Then, the two triangles $A_1A_2A_3$ and $B_1B_2B_3$ are perspective from point S, and by Desargues' theorem, the three points of intersection of pairs of corresponding sides are collinear. Since S_1 and S_2 are already known to be two of these three points, the corresponding sides A_1A_3 and B_1B_3 intersect in the third point S' of line S_1S_2.

Similarly, if C_1 and C_3 are another pair of corresponding points on L_1 and L_3, line C_1C_3 passes through S'. Thus, all lines joining corresponding points pass through S'. Therefore,

$$(A_1,B_1,C_1, \ldots) \overset{S'}{\overline{\overline{\wedge}}} (A_3,B_3,C_3, \ldots)$$

and the theorem is proved.

FIG. 7.10

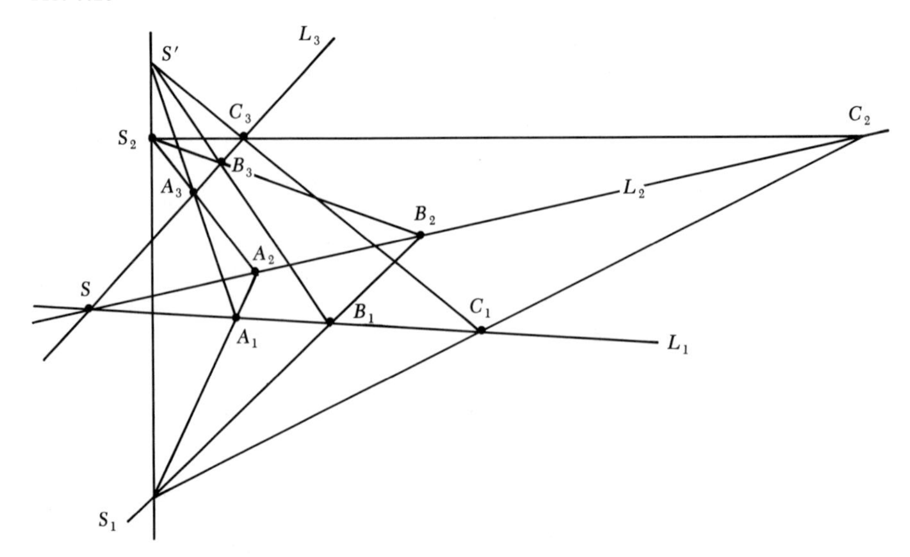

A statement and proof of the dual theorem is left as an exercise (see Exercise 7.2, question 2).

Finding a sufficient condition that a projectivity be a perspectivity, as was done in Theorem 7.2, is an accomplishment; but can a condition be found which is both necessary and sufficient? To be more explicit, an answer is now sought to the second question:

Question 2 *What is a necessary and sufficient condition that a projectivity be a perspectivity?*

A *necessary* condition is already at hand. If a projectivity

$$P_1,Q_1,R_1 \barwedge P_3,Q_3,R_3$$

between lines L_1 and L_3 (Fig. 7.11) is a perspectivity, the point of meeting, P_{13}, must be a self-corresponding point.

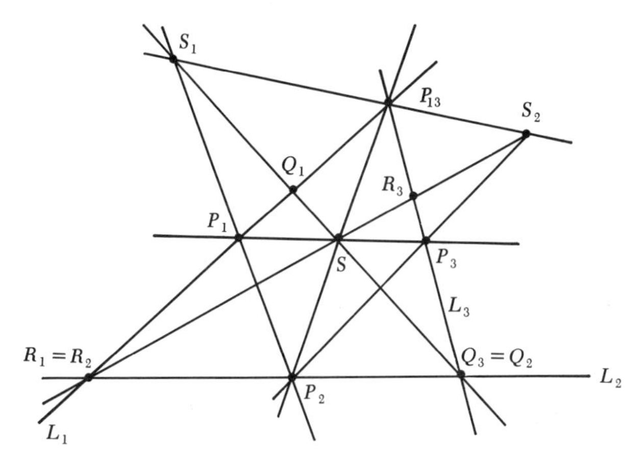

FIG. 7.11

To see if this necessary condition is also *sufficient*, let L_2 be the line determined by points R_1 and Q_3. It is noted first that the point P_{13} of intersection of lines L_1 and L_3 is self-corresponding if and only if it lies on the line S_1S_2, where S_1 is the center of a perspectivity from L_1 to L_2, and S_2 the center of perspectivity from L_2 to L_3.

Now, if the projectivity from L_1 to L_3 is to reduce to a perspectivity, its center S must be the point of meeting of lines R_1R_3 and Q_1Q_3, and P_1P_3 must also pass through S. On the assumption that such is the case, the projectivity is a perspectivity, and proof has been given of the theorem:

Theorem 7.4 *A necessary and sufficient condition that a projectivity between two distinct coplanar pencils of points (i.e., lines) of the plane be a perspectivity is that their point of intersection be self-corresponding.*

From the principle of duality follows the dual theorem:

Theorem 7.5 *A necessary and sufficient condition that a projectivity between two distinct coplanar pencils of rays be a perspectivity is that their common ray be self-corresponding.*

Exercise 7.2

1 If points A, B, C of a line L and points A', B', C' of a coplanar line L' (Fig. 7.12) are such that $A, B, C \barwedge A', B', C'$, explain how to locate two centers of perspectivity S and S' such that

$$A, B, C \overset{S}{\doublebarwedge} A', B'', C'' \overset{S'}{\doublebarwedge} A', B', C'$$

where B'', C'' are points of an auxiliary line L''. *Hint:* Take the auxiliary line L'' through a point A' of line L' intersecting lines SB and SC in points B'', C''.

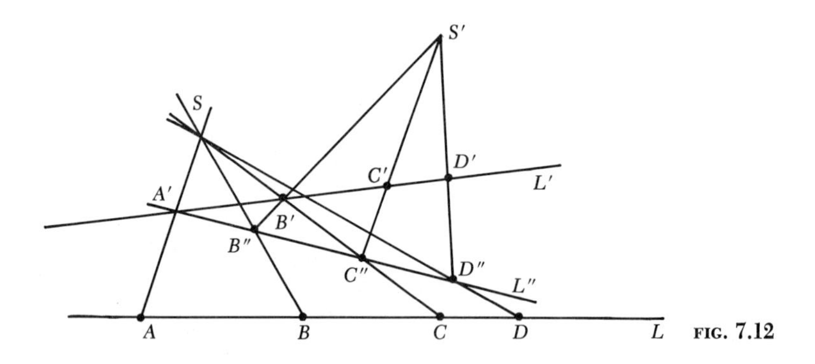

FIG. 7.12

2 Without using the principle of duality, prove the dual of Theorem 7.2, of Theorem 7.3, and of Theorem 7.4.

3 Prove that for any three collinear points A, B, C,

$$(A,B,C) \barwedge (A,C,B) \barwedge (B,A,C) \barwedge (B,C,A) \barwedge (C,A,B) \barwedge (C,B,A)$$

See ([100] p. 26).

4 Prove that for any four collinear points A, B, C, D,

$$(A,B,C,D) \barwedge (B,A,D,C) \barwedge (C,D,A,B) \barwedge (D,C,B,A)$$

5 Using the answer to question 4, show that

$$(A,B,C,D) \barwedge (B,A,C,D) \barwedge (C,D,B,A) \barwedge (D,C,A,B)$$

and hence show that in a harmonic set of points $H(AB,CD)$ each member of one pair is the harmonic conjugate of the other member with respect to the other pair.

6 If a line is set into correspondence with itself, A to A', B to B', C to C', etc., by means of two perspectivities, how many self-corresponding points are possible? Why? See ([100] p. 27).

7.5 Pappus' theorem

The new assumption used in proving Theorem 7.4 will be put in a more familiar form. If the collinear points S_1, P_{13}, S_2 of Fig. 7.11 are replaced by the points A, B, C, and the points R_2, P_2, Q_2 by the points A', B', C' (Fig. 7.13), the new assumption, here called Axiom 8′, reads as follows:

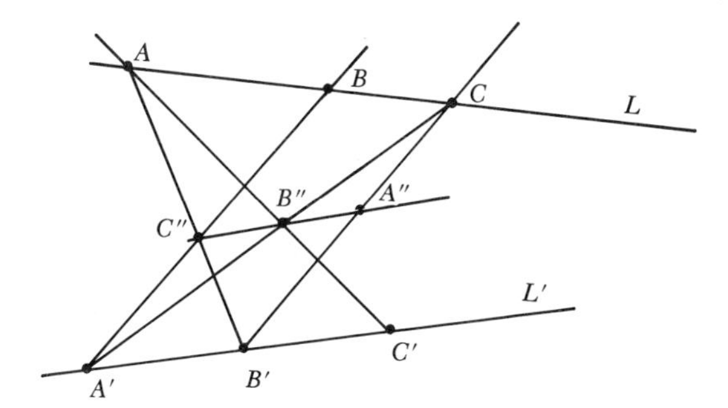

FIG. 7.13

Axiom 8′ *If A, B, C are any three distinct points of a line L, and A′, B′, C′ any three distinct points of a coplanar line $L' \neq L$, the three points*

$$A'' = \binom{BC'}{B'C} \qquad B'' = \binom{CA'}{C'A} \qquad C'' = \binom{AB'}{A'B}$$

are collinear.

The new axiom is seen to be the ancient Euclidean theorem discovered and proved by Pappus of Alexandria, and named in his honor "Pappus' Theorem." (See Exercise 4.1, question 6.)

The statement (Axiom 8′) cannot be proved on the basis of the present axiomatic structure. However, it is worthy of note that when no assumption is made about points outside the plane, Desargues' theorem for the plane may be proved by using Axioms 1 to 6 of Sec. 5.1 and Pappus' theorem taken as an axiom. (See question 3 of Exercise 7.3.)

Exercise 7.3

1 When the six points A, B, C of line L and A', B', C' of line L' (Fig. 7.13) are taken to be the vertices of a degenerate hexagon, the line of the points A'', B'', C'' is called a Pascal line of the hexagon. How many different hexagons may be formed from the six points by joining them in all possible orders? Does each one determine a Pascal line? Why?

2 a. Draw a figure illustrating the plane dual of Pappus' theorem: "If a, b, c are any three distinct lines through a point P and a', b', c' any three distinct lines through a point $P' \neq P$, then the three lines $(bc',b'c)$, $(ca',c'a)$, $(ab',a'b)$ are concurrent."

b. Prove the theorem using Axiom 8′.

3 Prove Desargues' theorem for the plane, using Axioms 1 to 6 of Sec. 5.1, Pappus' theorem taken as an axiom, and the assumption: "All points are in the same plane." See ([80] p. 63).

7.6 The fundamental theorem

It was shown in Theorem 7.1 that any three points A, B, C of a line L may be projectively related to any three points A', B', C' of a line $L' \neq L$ by *two* perspectivities. An intermediate line L'' was used in the process. If now a fourth point is selected on line L, its corresponding point D' on L' is automatically determined by the projectivities which related the corresponding points, and a natural question to ask is whether or not a different choice for the intermediate line L'' would determine a corresponding point on L'. In other words, is a *projectivity completely determined when three pairs of corresponding points are given?* The question is answered in the affirmative in the next theorem, called the fundamental theorem of projective geometry.

Theorem 7.6 *A projectivity between two coplanar pencils of points (lines of the plane) is uniquely determined when three pairs of corresponding points are given.*

proof: Again let S_1 and S_2 (Fig. 7.14) be the two centers of perspectivity establishing the projectivity

$$ABC \barwedge A'B'C'$$

between points A, B, C of line L and points A', B', C' of line $L' \neq L$.

The auxiliary line L'', used in the process, is drawn through a pair of noncorresponding points, say A and C' of lines L and L', and B'' is the projection of point B from S_1 onto L''.

If now a fourth point D of L projects from S_1 into D'' on L'' and D'' in turn projects from S_2 into point D' on L', the theorem will be proved by showing

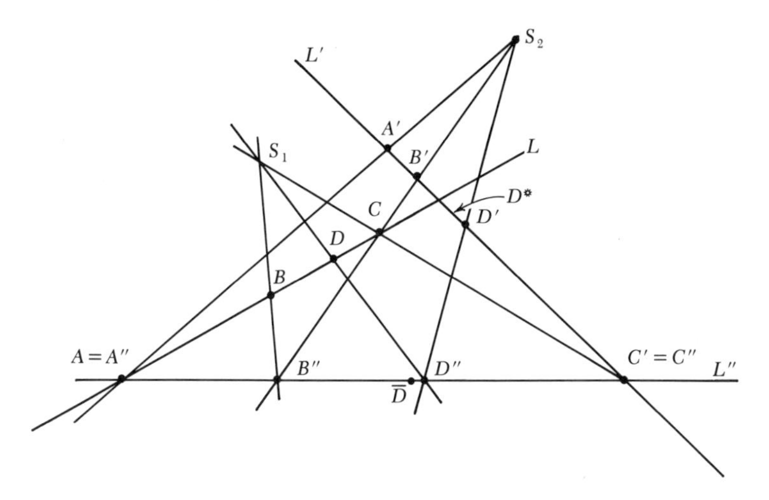

FIG. 7.14

that D' is independent of the sequence of perspectivities establishing the projectivity

$$A,B,C,D \barwedge A',B',C',D' \tag{1}$$

Suppose by a different sequence of perspectivities

$$A,B,C,D \barwedge A',B',C',D^* \tag{2}$$

where $D^* \neq D'$. If the projection from S_2 of D^* onto L'' is $\overline{D}$,

$$A',B',C',D^* \overset{S_2}{\overline{\barwedge}} A,B'',C',\overline{D} \tag{3}$$

then from (1) and (2)

$$A,B,C,D \barwedge A,B'',C',\overline{D}$$

But this last projectivity has A as a self-corresponding point and hence by Theorem 7.4 is a perspectivity, and $\overline{D}D$ passes through S_1. But S_1D intersects L'' in D'', and hence D'' and $\overline{D}$ coincide. Consequently, D^* coincides with D', and the theorem is proved.

Corollary *If a projectivity between two coplanar pencils of points (lines) of the plane leaves three distinct points of the pencil fixed, it leaves fixed every point of the pencil.*

The principle of duality is used to prove the dual theorem:

Theorem 7.7 *A projectivity between two pencils of rays of the plane is uniquely determined when three pairs of corresponding rays are given.*

Corollary *If a projectivity between two pencils of rays of the plane leaves three distinct rays of the pencil fixed, it leaves fixed every ray of the pencil.*

7.7 Concerning the equivalence of Pappus' theorem and the fundamental theorem

Pappus' theorem (Axiom 8′) was used in proving the fundamental theorem. The equivalence of these two theorems is established now by showing that Pappus' theorem is a consequence of the fundamental theorem.

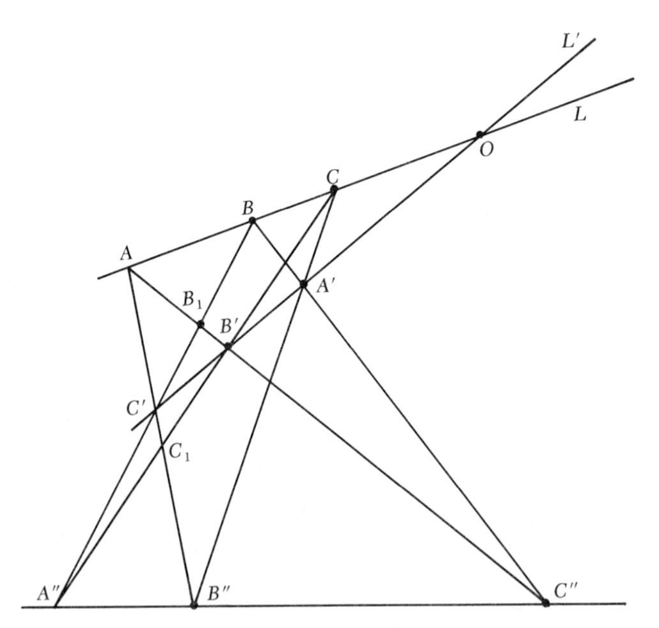

FIG. 7.15

Suppose in Fig. 7.15,

$$B_1 = \binom{BC'}{AB'} \qquad C_1 = \binom{C'A}{B'C}$$

where A, B, C are points of line L, and A', B', C' of line L' intersecting L at O. Let A'', B'', C'' be the points

$$A'' = \binom{BC'}{B'C} \qquad B'' = \binom{AC'}{A'C} \qquad C'' = \binom{AB'}{A'B}$$

Since C'', B', B_1, A are collinear,

$$(C'',B',B_1,A) \overset{B}{\overline{\wedge}} (A',B',C',O)$$

But

$$(A',B',C',O) \overset{C}{\overline{\wedge}} (B'',C_1,C',A)$$

and therefore

$$(C'',B',B_1,A) \overline{\wedge} (B'',C_1,C',A)$$

But in this latter projectivity, A is a self-corresponding point, and hence by Theorem 7.4, the projectivity is a perspectivity. The lines

$$B''C'' \qquad B'C_1\ (= B'C) \qquad B_1C'\ (= BC')$$

are therefore concurrent at the point A'' of meeting of the lines $B'C$ and BC', and the three points A'', B'', C'' are therefore collinear.

Exercise 7.4

1 A line L may be set into correspondence with itself as follows: Let E, F be any two points of L, and let L' be a line through one of these points, say E. On L' select two points S_1 and S_2 distinct from E, and locate the harmonic conjugate G of E with respect to S_1 and S_2 (Fig. 7.16). Then project points A, B, C of L through S_1 into points A'', B'', C'' of the line FG and the latter points, in turn, from S_2 back to points A', B', C' of the original line L.

Since

$$A,B,C \overset{S_1}{\overline{\wedge}} A'',B'',C'' \overset{S_2}{\overline{\wedge}} A',B',C'$$

then

$$A,B,C \overline{\wedge} A',B',C'$$

What points of L are invariant under the projectivity determined by the three points A, B, C and their corresponding points A', B', C'?

2 Will the above projectivity send point A' into A? (If it does, the projectivity is called an involution.) *Hint:* See ([100] p. 27).

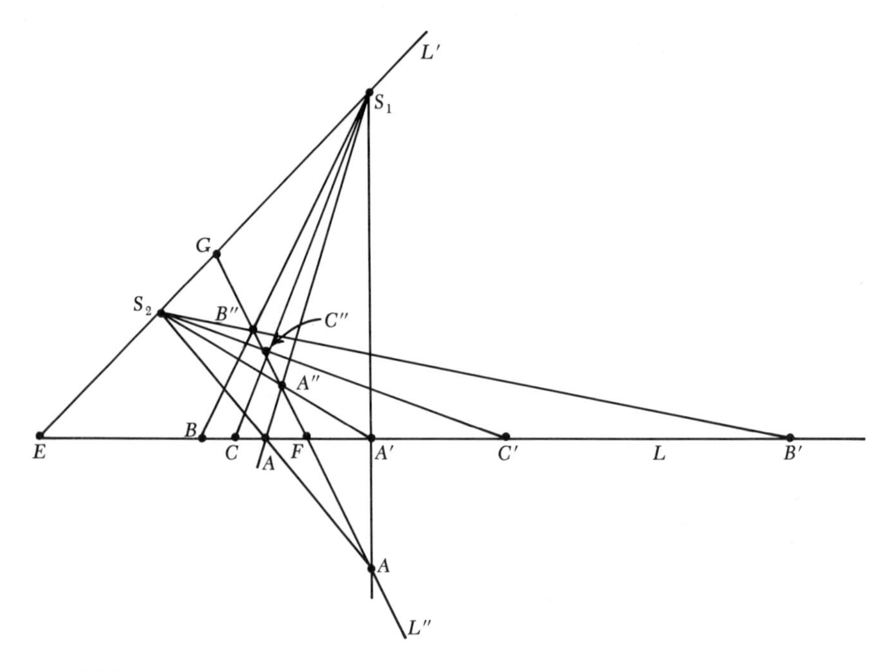

FIG. 7.16

PROJECTIVE THEORY OF CONICS

Approximately two thousand years after the Greeks developed their extensive metric theory of conics, projective nonmetric properties of these curves were discovered. The beauty, simplicity, and elegance of these new properties have justly earned for themselves a place in the elementary curriculum along with the highly practical metric theory. Among other things this theory enabled Kepler in 1609 to announce his famous law that the earth and other planets moved in elliptical orbits about the sun. This law in turn paved the way for the even greater discovery of Newton's law of gravitation, which relies upon the fact that planetary orbits are ellipses.

There is no denying the importance of these metric properties, but neither can one minimize the importance of the nonmetric properties. They are of especial interest and appeal to the artist, the draftsman, the architect, and the engineer.

Perspectivities and projectivities figure prominently in the following introduction to the modern and stimulating nonmetric theory of conics. Unfortunately, there is time here to stress only the initial stages in this marvelous new development.

7.8 Corresponding elements in two projective, nonperspective forms

A method will be explained first for locating corresponding elements in two projective nonperspective pencils of points.

By Theorem 7.6, not more than three points in one pencil and their corresponding points in another may be selected at random.

Suppose that the pencil of points on line L (Fig. 7.17) is projective (but not perspective) with a pencil of points on a second line L' in such a way that the points A, B, C of line L correspond, respectively, to the points A', B', C' of line L'. It is desired to find the point D' of line L' corresponding to a fourth point D of line L.

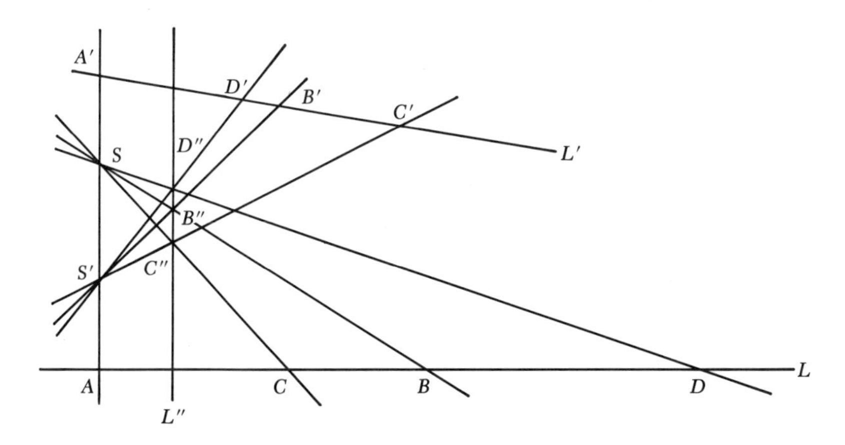

FIG. 7.17

On any line, say AA', joining a pair of corresponding points, choose two points S and S', and join S to points of L and S' to points of L'. In the pencils of rays at S and S', ray SS' is self-corresponding. Hence by Theorem 7.4 these pencils are perspective, and all pairs of corresponding rays will intersect in points of a straight line. But, the intersection B'' of the corresponding rays SB, $S'B'$ and the intersection C'' of the corresponding rays SC, $S'C'$ determine the line L'', on which pairs of corresponding rays intersect. If then ray SD meets L'' in a point D'', rays $S'D''$ and SD are corresponding, and line $S'D''$ meets line L' in the required point D'.

Dually, suppose that in the projective pencils of lines with vertices at S and S' (Fig. 7.18), rays a, b, c of the pencil at S correspond respectively to rays a', b', c' of the pencil at S' and that it is desired to find the ray d' of pencil S' corresponding to a fourth ray d of the pencil at S.

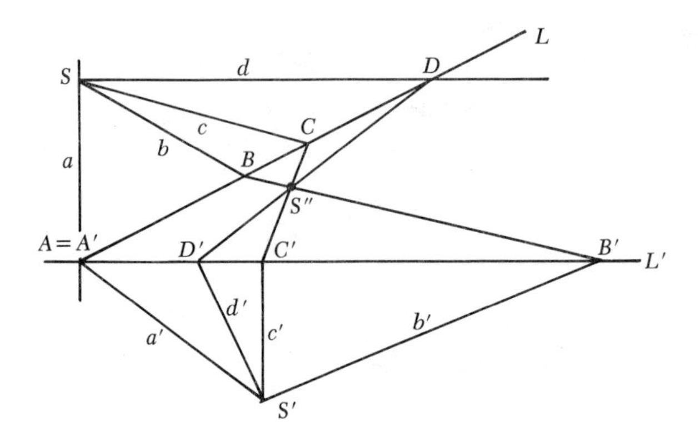

FIG. 7.18

Draw two lines L and L' through a point, say A, of meeting of corresponding rays, and let rays a, b, c meet L in the points A, B, C, and let rays a', b', c' meet L' in the points A, B', C' respectively. Since the two pencils of points on L and L' have point A as a self-corresponding point, these two pencils are perspective, and lines joining corresponding points pass through the point S'' of meeting of the lines BB' and CC'. If then ray d meets line L in point D, the line $S''D$ meets line L' in a point D' such that $S'D' = d'$ is the required ray.

Exercise 7.5

1 Show that the position of D' (Fig. 7.17) is independent of the choice of the points S and S' on line AA'. *Hint:* See ([61] p. 67).

2 Show that the position of ray d' (Fig. 7.18) is independent of the choice of lines L and L' through the point A.

7.9 Point and line conics

It will be recalled that in metric geometry, a conic is defined as the locus of a point P which moves so that its distance from a fixed point, called the focus, and its distance from a fixed line, called the directrix, are in a constant ratio. Values of this ratio are then used to classify the resulting conics as ellipses, hyperbolas, or parabolas.

Note the absence of metric concepts in the following *projective definitions* of point and line conics.

Definition 1 *A point conic consists of the points of intersection of corresponding rays in two projective nonperspective pencils.*

Definition 2 *A line conic is the envelope of lines joining corresponding points in two projective nonperspective pencils of points.*

The definitions are illustrated by figures in the Euclidean plane, where metric concepts are available for describing the generating pencils.

In Fig. 7.19 the projective pencils at S and S' are congruent, i.e., the angle between a pair of rays a, b at S equals the angle between their corresponding rays a', b' at S'. As ray a rotates about S in the counterclockwise direction, ray a' rotates about S' in the same direction, and the point A of intersection of corresponding rays moves on a circle.

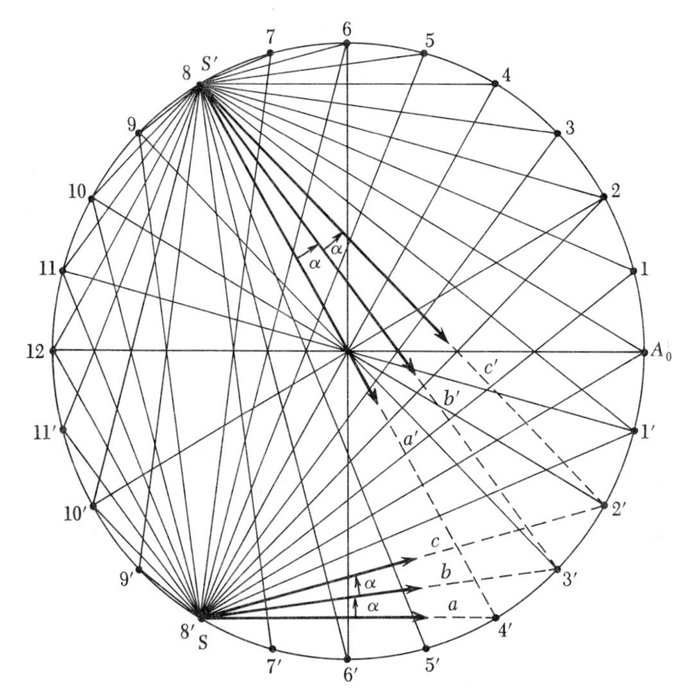

FIG. 7.19

Now, a quite remarkable fact is observed. If the rotations at S and S' are in opposite directions, the point A no longer moves on a circle but on a hyperbola (Fig. 7.20).

Line conics are illustrated in Figs. 7.21 and 7.22.

In Fig. 7.21 lines AA', BB', CC' joining corresponding points in two projective pencils of points on lines L and L' are tangent to an ellipse, and in Fig. 7.22 to a parabola.

In nonmetric projective geometry these conics cannot be classified, but they can be shown to have the following properties in common with the conics of metric geometry:

Property 1 *A line intersects a point conic in two points at most.*

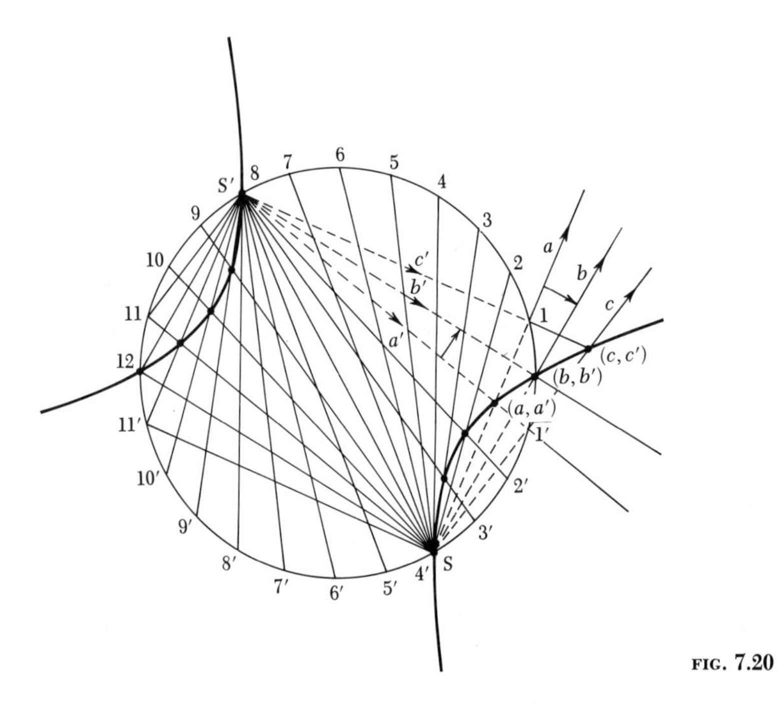

FIG. 7.20

If a line met the conic in three points, the projectivity would be a perspectivity, and points of intersection of corresponding rays would all lie on a line.

Property 2 *A point conic is determined by five points.*

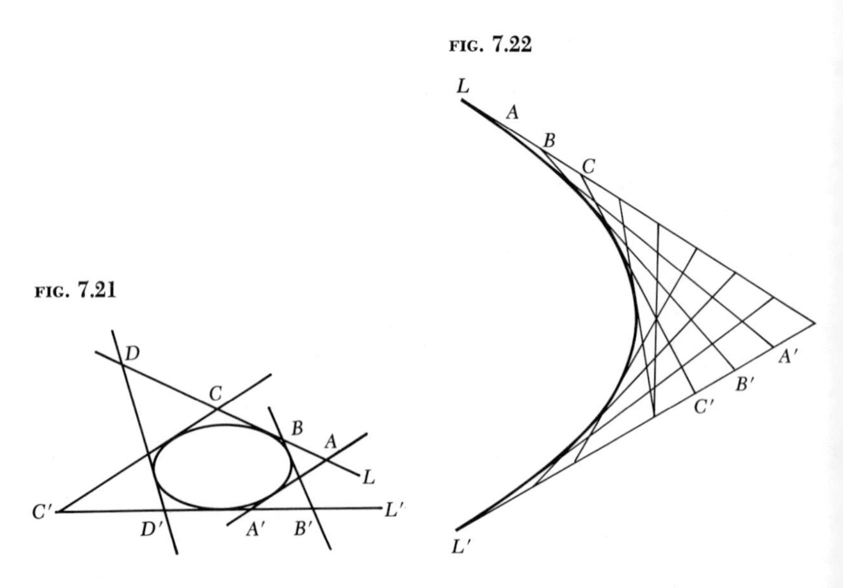

FIG. 7.22

FIG. 7.21

A conic passes through the three points $(a,a') = A$, $(b,b') = B$, $(c,c') = C$ of intersection of corresponding rays in two projective pencils at S_1 and S_2, and the conic is determined by the three rays a, b, c and their corresponding rays a', b', c'; also, the conic passes through the vertices S_1 and S_2, for a ray of the pencil at one vertex meets its corresponding ray at the other vertex.

Statement and proof of the dual properties are left as exercises.

Exercise 7.6

1 How may the projective pencils of points of Figs. 7.21 and 7.22 be distinguished so that in one case the envelope of lines AA', BB', etc., is an ellipse, and in the other a parabola?

2 Pairs of projectively related pencils of rays may be obtained by projecting all the points P of a straight line L from two different centers S and S''. Then, corresponding rays a, a'; b, b'; . . . intersect in points of L. Now take the pencil of rays at S'', and move it rigidly to a new position S'. Prove that the resulting pencil at S' is projective with the pencil at S.

3 Experiment with the drawing of point conics by starting with your own projective pencils of rays a, b, c at S and their corresponding rays a', b', c' at S', and show in a drawing the resulting curves. Can you predict in advance what the curve will be? Explain.

4 In Fig. 7.20, how many pairs of corresponding lines are parallel in the two projective pencils generating the conic? Give reasons for your answer.

7.10 Pascal's and Brianchon's theorems

Two famous theorems of projective geometry are Pascal's theorem (Theorem 7.8) and its dual Brianchon's theorem (Theorem 7.9) named after their respective discoverers, Pascal and Brianchon. Although Pascal discovered his theorem in the year 1640, its dual was not discovered until the year 1806.

Theorem 7.8 *The points of intersection of the opposite sides of a hexagon inscribed in a conic are collinear.*

Pascal's theorem is proved when A, C_1 are vertices of the pencils determining the conic and A,B', C,A', B,C' are vertices of the inscribed hexagon. If A'', B'', C'' are the points

$$A'' = \begin{pmatrix} BC' \\ B'C \end{pmatrix} \qquad B'' = \begin{pmatrix} AC' \\ A'C \end{pmatrix} \qquad C'' = \begin{pmatrix} AB' \\ A'B \end{pmatrix}$$

the theorem is proved by showing that A'', B'', C'' are collinear.

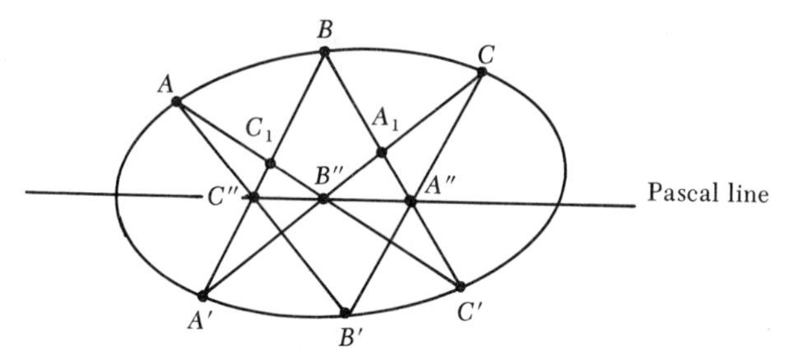

FIG. 7.23

Let C_1 and A_1 be the points

$$C_1 = \begin{pmatrix} AC' \\ BA' \end{pmatrix} \qquad A_1 = \begin{pmatrix} BC' \\ A'C \end{pmatrix}$$

Then, because the pencils at A and C are projective,

$$(BA'C''C_1) \barwedge (BA_1A''C')$$

But, in this projectivity, B is a self-corresponding point, and hence by Theorem 7.5 the projectivity is a perspectivity.

The lines $A'A_1$, $C''A''$, C_1C' joining corresponding points are, therefore, concurrent at the point B'' of meeting of lines AC' and $A'C$, and the three points A'', B'', C'' are collinear. Thus the theorem is proved.

The line of the points A'', B'', C'' is called the Pascal *line of the hexagon.*

When the vertices A, B, C of the hexagon are on a line and the remaining vertices A', B', C' are on a second line (i.e., the conic is degenerate), Pascal's theorem reduces to Pappus' theorem.

Brianchon's theorem, which follows from Pascal's by the principle of duality, is stated next.

Theorem 7.9 *The lines joining the opposite vertices of a hexagon circumscribed about a conic are concurrent.*

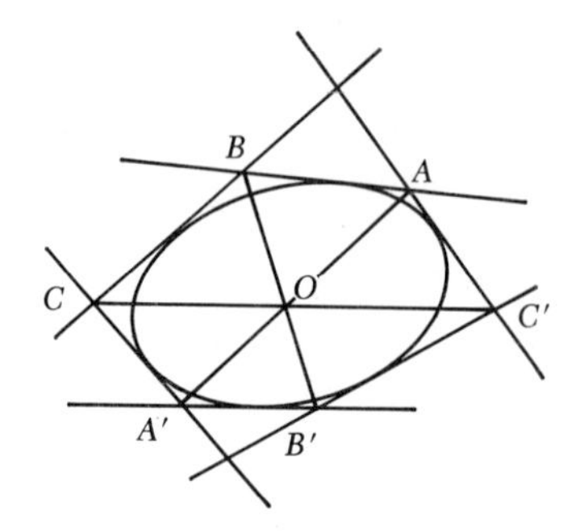

FIG. 7.24

The theorem is illustrated in Fig. 7.24, where lines AA', BB', CC', joining opposite vertices of the hexagon, are concurrent at O.

PROBLEMS FOR SPECIAL STUDY

Exercise 7.7

1 Given five points of a conic and an arbitrary line through one of them, find where the line cuts the curve a second time. *Hint:* Let A, B, C, D, E (Fig. 7.25) be five given points of the required curve, and let X be the arbitrary line through point A. If P is the sixth point of the curve through these five points, then in the hexagon $ABCDEP$ take as pairs of opposite sides AB and DE, BC and EP, CD and PA, which is the given line X, and use Pascal's theorem to locate the point P.

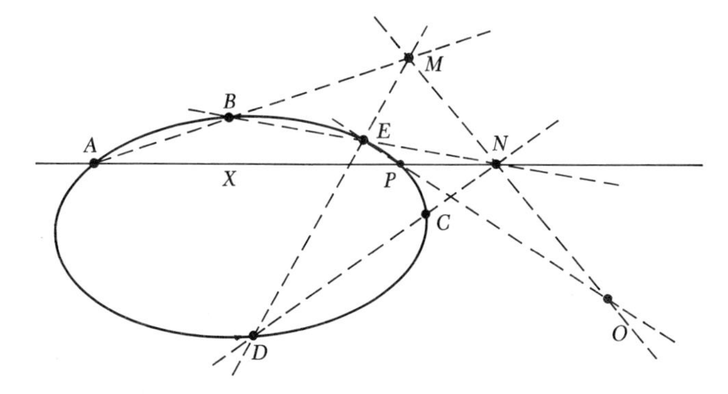

FIG. 7.25

2 A conic is given by five of its tangents. Construct the point of contact on one of these tangents.

3 Show that six points on a conic determine 60 Pascal lines.

4 Show that Pappus' theorem implies the validity of Desargues' theorem for the plane. See ([113] Chap. 4).

5 Show that if a quadrangle is inscribed in a conic, the tangents at its vertices meet in pairs on the sides of the diagonal point triangle.

Concluding remarks

Much of the discussion of this chapter has centered around perspectivities and projectivities. The fundamental theorem of projective geometry has been proved and the significance of Pappus' theorem pointed out. Pascal's and Brianchon's theorems have been noted, and a brief introduction has been given to the projective (nonmetric) theory of conics.

A climax, so to speak, has been reached in the presentation of the synthetic (pure) projective geometry of the text. Even though the material could be magnified tenfold, more will be gained now by a shift of emphasis from pure to analytic projective geometry. In this modern development is found the algebraic theory that unifies the entire elementary field and links geometry with a multitude of highly abstract modern concepts. Linear algebra, a relatively recent mathematical development of wide applicability, combines a variety of the algebraic and geometric concepts that are introduced later in the text.

PART THREE

ALGEBRAIC PROJECTIVE GEOMETRY AND LINEAR ALGEBRA

8 natural homogeneous coordinates: matrices

In any survey of the field of important developments in geometry, there is one subject without which no introduction to the science is complete. It is coordinate geometry, by which every geometric object and every geometric operation may be referred to the realm of numbers.

Decisive steps in the arithmetization of geometry were taken as early as 1629 by Fermat, but chief credit for this new development goes to the Frenchman Descartes, who, in 1637, published his "La Géométrie," which linked geometry with algebra and the number system and gave to the world the powerful tools needed in later developments.

Opposition to the new geometry was almost immediate, and before long, mathematicians, like politicians, were divided into rival parties, one called purists, and the other nonpurists. The purists argued that the introduction of number concepts debased an otherwise pure science. Descartes and his followers, the nonpurists, were adamant. They argued that figures and geometric intuition should be replaced by number concepts and algebraic theory.

The clash between the rival factions was long and bitter, but when the tumult finally quieted down, each side was found to have made concessions to the other. The purists, in their strict avoidance of the new techniques, often encountered problems in which some algebraic formulation was unavoidable. Nonpurists, on the other hand, often lost themselves in long, tiresome manipulations. It was, therefore, inevitable that the two viewpoints would merge.

Today, coordinate geometry and algebra take over when figures, synthetic methods, and geometric intuition fail. This fortunate turn of events is the product of a long, gradual historical growth that has greatly enlarged the scope of classical theory.

It is assumed in what follows that the reader is already familiar with the elementary coordinate geometry of the plane and 3-space, where to a point of the line there corresponds a coordinate number x, to a point of the plane a pair of coordinate numbers (x,y), and to a point of space the coordinate number triple (x,y,z), called in each case the nonhomogeneous coordinates of the point in question. (See [116].)

In projective geometry, it was von Staudt's algebra of throws which associated a real number x with each point of the line. Considerably later, the theory was extended to include the case when the coordinate x is a complex number. In current usage, *according as a coordinate geometry is based on the real or the complex number system, it is called, respectively, real or complex coordinate geometry.*

After the reader has been introduced to both systems, he will be in a better position to compare, evaluate, and appreciate the advantages of each. Certainly, a full appreciation of either geometry involves a knowledge of the other.

Attention is being directed first to *real* projective coordinate geometry and basic concepts of linear algebra. This union of geometric and algebraic theory is named in today's literature "algebraic" geometry.

8.1 A Euclidean model of the projective plane

The study of coordinates and algebraic methods in the projective plane begins with the construction of an analytic model in which each point is represented by a triple of real numbers called *homogeneous* coordinates of the point. A Euclidean model to be used in constructing the analytic model will be described first.

In a Euclidean 3-space, consider the set of all lines through a point O of this space. Each pair of distinct lines of this set will determine a plane. The resulting set of planes, together with the totality of lines through O, constitutes a model of the projective plane. To see why, first rename these lines and planes of space, calling the planes "lines" and the lines "points". Let the relation of a point being on a line be taken to mean that a line lies in a plane, as shown in the table below:

Euclidean 3-space	*Projective plane*
Lines through O	Points
Planes through O	Lines
Line in a plane	Point on a line

It can then be shown that *lines* and *planes* through O satisfy the incidence and existence axioms for *points* and *lines* of the projective plane. This is accomplished by simply recalling known properties of Euclidean 3-space. The verification is made in the parallel columns below.

Incidence Axioms

Euclidean 3-space	*Projective plane*
1 If a and b are distinct lines through point O, there is at least one plane through O on both a and b.	*1'* If A and B are distinct points, there is at least one line on both A and B.
2 If a and b are distinct lines through point O, there is not more than one plane through O on both a and b.	*2'* If A and B are distinct points, there is not more than one line on both A and B.
3 If a plane through the origin intersects two faces of a trihedral angle whose vertex is the origin, then it intersects the third face.	*3'* If a line intersects two sides of a triangle, then it intersects the third side.

Existence Axioms

4 There exists at least one plane through point O.	*4'* There exists at least one line.
5 There are at least three lines on every plane through O.	*5'* There are at least three distinct points on every line.
6 Not all lines through O are on the same plane through O.	*6'* Not all points are on the same line.
7 Every plane through O is in Euclidean 3-space.	*7'* Every line is in the plane.

8.2 An analytic model. Homogeneous point coordinates

Now that points and lines of the projective plane have been represented by lines and planes of Euclidean 3-space, the analytic model may be described.

In Euclidean 3-space, let three mutually perpendicular planes x_2x_3, x_1x_3, x_1x_2 (Fig. 8.1) intersect in a point O. Then to each point P of this space corresponds the real number triple (x_1,x_2,x_3), where x_1, x_2, x_3 are the respective signed distances of P from these three planes.

Consider any line through the origin O of this coordinate system. Such a line is determined by a set of three direction numbers $a_1, a_2, a_3 \neq 0, 0, 0$. Any other set (ta_1,ta_2,ta_3) determines the same line, where t is any real number $\neq 0$.

The number triple (a_1,a_2,a_3) is said to be equivalent to another number triple (b_1,b_2,b_3) if there exists a number $k \neq 0$ such that

$$b_1 = ka_1 \qquad b_2 = ka_2 \qquad b_3 = ka_3$$

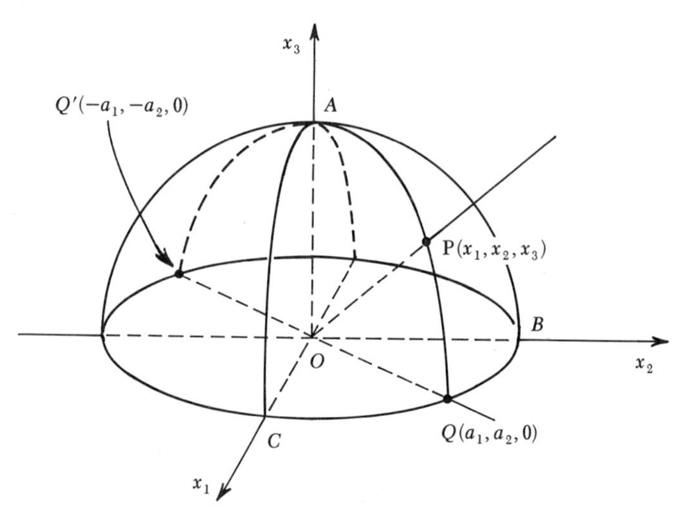

FIG. 8.1

The set of all mutually equivalent number triples forms an equivalence class which is denoted simply $\{a_1,a_2,a_3\}$. Since there exists a 1-1 correspondence between the equivalence classes $\{x_1,x_2,x_3\}$, not all of whose elements are zero, and the lines through the origin of Euclidean 3-space, and since each such line is a model of a point of projective plane, *these equivalence classes constitute an analytic model of points of the projective plane.*

Any representative, such as (a_1,a_2,a_3), of a particular class will be called *homogeneous coordinates of a point of the real projective plane* if the number triple is proportional to a triple of real numbers. Thus (1,2,2) and $(i,2i,2i)$, where $i = \sqrt{-1}$, represent the same real point.

The picture becomes clearer if one further step is taken. Consider the hemisphere of unit radius lying above the x_1x_2 plane in Euclidean 3-space (Fig. 8.1). If a line through the origin does not lie in this plane, it cuts the hemisphere in a unique point $P(x_1,x_2,x_3)$, where

$$x_1^2 + x_2^2 + x_3^2 = 1$$

Furthermore, the coordinates of point P are a set of direction numbers for the line OP and hence may be used as homogeneous coordinates of the point.

One slightly disturbing feature, however, is present. A line through the origin, lying in the x_1x_2 plane, cuts the hemisphere in two points $Q(a_1,a_2,0)$ and $Q'(-a_1,-a_2,0)$. Since both of these number triples are direction numbers of the line, either one may be used as homogeneous coordinates of the point Q or Q'. To avoid having two points of the hemisphere correspond to one point of the projective plane, these two points Q and Q' are now identified. With this agreement, the number triples $(a_1,a_2,0)$ and $(-a_1,-a_2,0)$ represent the same point. Points of the unit hemisphere in Euclidean 3-space then constitute another model of points of the projective plane, and Euclidean coordinates of these points may, therefore, be taken as homogeneous coordinates of points of the projective plane.

Exercise 8.1

1 Locate on the hemisphere of Fig. 8.1 points corresponding to the points $B(1,0,0)$, $C(-1,0,0)$, $D(0,1,0)$, $E(0,0,1)$, $F(1,2,3)$, $G(-1,-2,-3)$.

2 If a, b are any real numbers, the number triple $(a,1,0)$ represents a *set* of points. Describe their position on the hemisphere. Do the same for the points $(1,a,0)$ and $(a,b,0)$, $a, b \neq 0, 0$.

8.3 Lines in the analytic model. Homogeneous line coordinates

As shown in Sec. 8.1, a projective line corresponds to a plane through the origin of Euclidean 3-space. The equation of such a plane is considered next. It is known to be

$$d_1x_1 + d_2x_2 + d_3x_3 = 0 \tag{1}$$

where d_1, d_2, d_3 are proportional to a set of real constants $\neq 0, 0, 0$.

Since a point which satisfies (1) also satisfies the equation

$$kd_1x_1 + kd_2x_2 + kd_3x_3 = 0 \tag{2}$$

where k is any number $\neq 0$, the plane is determined either by the real number triple (d_1,d_2,d_3), not all of whose elements are zero, or by any other number triple (kd_1,kd_2,kd_3) where k is a constant $\neq 0$. A 1-1 correspondence therefore exists between the equivalence classes $\{d_1,d_2,d_3\} \neq 0, 0, 0$ and planes through the origin and hence also between these equivalence classes and lines of the projective plane.

Any representative (d_1,d_2,d_3) of the equivalence class $\{d_1,d_2,d_3\}$ will be called homogeneous line coordinates of a line of the real projective plane.

In the hemispherical model (Fig. 8.1), planes through the origin are seen to intersect the hemisphere in great semicircles whose end points are identified. These "closed semicircles" are, therefore, representatives of lines of the projective plane, the only exception being in the case of the x_1x_2 plane, which cuts the hemisphere in a circle. But, because diametrically opposite points are being identified on this boundary circle, the latter also represents a projective line, i.e., the one with homogeneous line coordinates $(0,0,1)$.

Other examples of line coordinates are $(1,0,0)$ and $(0,1,0)$, corresponding to the respective semicircles AC and AB of Fig. 8.1.

8.4 Incidence of points and lines. Duality

In the axiomatic definition of a projective plane, the concept of incidence of point and line was undefined. It was, however, required to satisfy certain conditions called incidence axioms. This same concept of incidence, in the case of the analytic model, takes a concrete form now to be explained.

The incidence of a point and a line follows from the fact that the corresponding line through the origin lies in the appropriate plane through the origin. Since the equation of such a plane is

$$d_1x_1 + d_2x_2 + d_3x_3 = 0 \tag{1}$$

and the parametric equations of such a line are

$$x_1 = a_1t \qquad x_2 = a_2t \qquad x_3 = a_3t \tag{2}$$

where a_1, a_2, a_3 are its direction cosines, the line will lie in the plane if its coordinates satisfy the equation of the plane for all $t \neq 0$, that is, if

$$d_1a_1 + d_2a_2 + d_3a_3 = 0 \tag{3}$$

Since the converse statement also holds, proof has been given of the following theorem:

Theorem 8.1 *A point with homogeneous coordinates (a_1,a_2,a_3) and a line with homogeneous coordinates (d_1,d_2,d_3) are incident if and only if*

$$d_1a_1 + d_2a_2 + d_3a_3 = 0$$

When the point has homogeneous coordinates (x_1,x_2,x_3), (3) takes the standard form

$$d_1x_1 + d_2x_2 + d_3x_3 = 0 \tag{4}$$

called the *point equation of the line.* Equation (4) is, therefore, the analytic representation of a pencil of points.

Now consider Eq. (4) from another point of view. This time suppose that the point (x_1,x_2,x_3) is fixed and the coefficients d_1, d_2, d_3 vary. Then, to each number triple (d_1,d_2,d_3) satisfying (4) there corresponds a line through the fixed point (x_1,x_2,x_3). Equation (4), therefore, represents a pencil of lines through this point.

It is well now to distinguish between number triples which represent points and those which represent lines. This will be done by using the number triple (x_1,x_2,x_3) to represent a variable point and the number triple $[u_1,u_2,u_3]$ to represent a variable line.

The linear equation

$$a_1u_1 + a_2u_2 + a_3u_3 = 0$$

is the *line equation of the fixed point* (a_1,a_2,a_3) in accordance with the definition:

> *A line equation of a point is an equation in variable line coordinates* $[u_1,u_2,u_3]$ *which is satisfied by the coordinates of only those lines which pass through the point.*

The line equation of the point (2,3,−4) is

$$2u_1 + 3u_2 - 4u_3 = 0$$

and line coordinates of two lines of this pencil are [1,2,2] and [3,2,3], as can be easily verified (see Exercise 8.2, questions 3a and b).

Since the equation

$$u_1x_1 + u_2x_2 + u_3x_3 = 0$$

represents a point, if the x's are fixed and the u's are variables, or a line, if the u's are fixed and the x's are variables, the statement

$$2 \cdot 3 + 3 \cdot 2 - 4 \cdot 3 = 0$$

may be taken to mean that the point (3,2,3) lies on the line [2,3,−4] or that the point (2,3,−4) lies on the line [3,2,3]. This symmetry is the basis of duality between point and line in analytic projective geometry.

Exercise 8.2

1 Locate on the hemisphere of Fig. 8.1 the semicircles corresponding to the line whose point equation is

a. $x_1 + x_2 + x_3 = 0$
b. $x_3 = 0$

2 What point of the hemisphere of Fig. 8.1 is represented by the line equation

a. $u_1 + u_2 + u_3 = 0$
b. $u_3 = 0$
c. $u_1 - u_2 = 0$

3 Find (a) the point equation of the line through the points $A(1,0,-2)$ and $B(2,3,4)$, and (b) the point of intersection of the lines $a[1,0,-2]$ and $b[2,3,4]$.

4 Show that the points $A(a_1,a_2,a_3)$ and $B(b_1,b_2,b_3)$ are not distinct if there exists constants $(k, l) \neq (0, 0)$ such that

$$ka_i + lb_i = 0 \qquad i = 1, 2, 3$$

8.5 Concerning the relation of nonhomogeneous to natural homogeneous coordinates

The simple substitution

$$x = \frac{x_1}{x_3} \qquad y = \frac{x_2}{x_3} \qquad x_3 \neq 0 \tag{1}$$

establishes a correspondence between ordinary points $P(x_1,x_2,x_3)$ of the projective plane and points $P(x,y)$ of the Euclidean plane, where x, y are the respective distances of P from the origin O of a rectangular coordinate system. The numbers x, y are then called nonhomogeneous coordinates of the projective point $P(x_1,x_2,x_3)$. See Fig. 8.2.

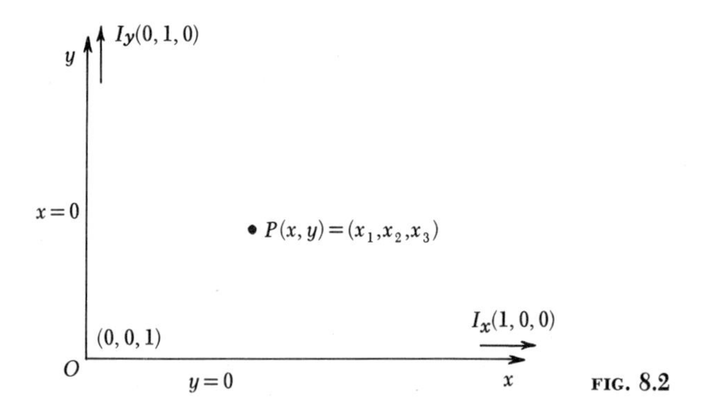

FIG. 8.2

There is no Euclidean point corresponding to an ideal point $(x_1,x_2,0)$, but there is some useful terminology to describe ideal points and link them with Euclidean geometry. First it is noted that the equation

$$d_1x_1 + d_2x_2 + d_3x_3 = 0 \tag{2}$$

of a projective line becomes, under (1), the equation

$$d_1x + d_2y + d_3 = 0 \tag{3}$$

representing a line in the Euclidean plane with slope $-d_1/d_2$ if $d_2 \neq 0$ and a vertical line if $d_2 = 0$.

Since $(d_2, -d_1, 0)$ is the ideal point of (2), it is convenient to call this point the ideal point in the direction $-d_1/d_2$ or in the vertical direction, depending on whether d_2 is or is not equal to zero.

An ideal point is often referred to as an infinitely distant point, since as x_3 approaches zero, both x and y approach infinity. Thus the three particular points

$$O(0,0,1) \qquad I_x(1,0,0) \qquad I_y(0,1,0)$$

are pictured in Fig. 8.2 as the origin O of a rectangular coordinate system, the infinitely distant point $I_x(1,0,0)$ on the x axis, and the infinitely distant point $I_y(0,1,0)$ on the y axis.

The numbers x_1, x_2, x_3 satisfying (1) are called *natural homogeneous coordinates* and the triangle OI_yI_x is called the reference triangle for a *natural coordinate system* in the projective plane.

In the drawings which follow, ordinary points and lines are represented by their corresponding elements in the familiar x, y coordinate plane.

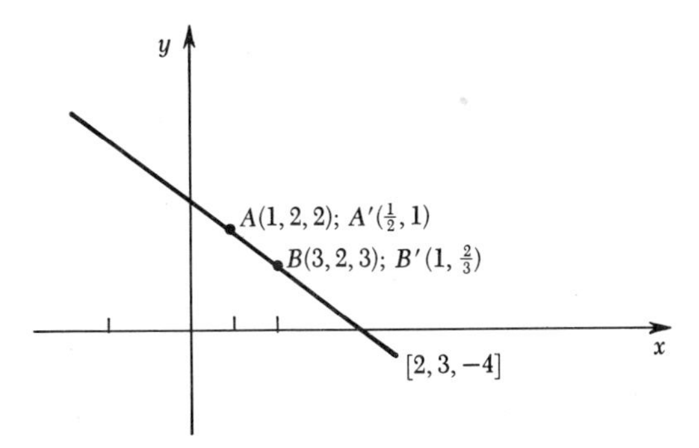

FIG. 8.3

For example, points $A(1,2,2)$ and $B(3,2,3)$ (Fig. 8.3) on the line

$$2x_1 + 3x_2 - 4x_3 = 0$$

are points $A'(1/2,1)$, $B'(1,2/3)$ of the line

$$2x + 3y - 4 = 0$$

Also, the two lines [3,2,3], [1,2,2] on the point $(2,3,-4)$ are the lines

$$3x + 2y + 3 = 0$$
$$x + 2y + 2 = 0$$

intersecting at the point $(-1/2,3/4)$ (Fig. 8.4).

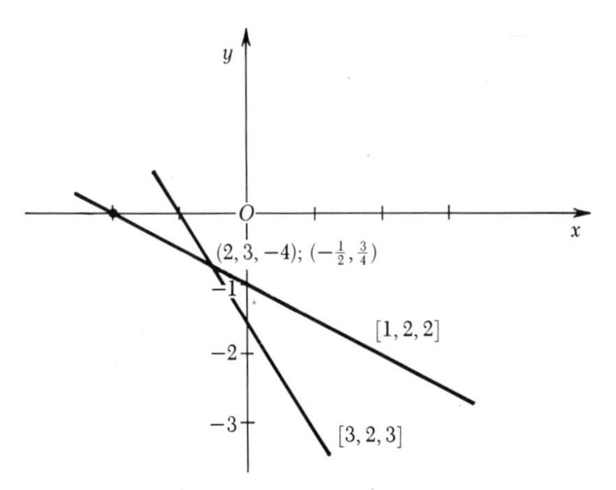

FIG. 8.4

Exercise 8.3

1 Give rectangular nonhomogeneous coordinates for points E, F, G of Exercise 8.1. Show them in an x, y rectangular coordinate system.

2 What figures are represented by the number triples (a) (1,2,3) and (b) [1,2,3]? Show both figures in a drawing.

3 Give a set of homogeneous coordinates for each of two points on the line [1,0, −1], and show them in a drawing.

4 Give a set of homogeneous coordinates for each of two lines on the point (1,0, −1), and show them in a drawing.

5 Find equations in homogeneous coordinates of the coordinate axes $x = 0$ and $y = 0$ of a rectangular coordinate system. Where in Fig. 8.2 is the line $x_3 = 0$?

8.6 Some elementary algebraic theory

A temporary digression is needed now to recall some algebraic theory pertinent to the discussion at hand.

It is known from elementary algebra that the system of *three* nonhomogeneous linear equations in the three unknowns x_1, x_2, x_3, that is,

$$\begin{aligned} a_{11}x_1 + a_{12}x_2 + a_{13}x_3 &= a_{14} \\ a_{21}x_1 + a_{22}x_2 + a_{23}x_3 &= a_{24} \\ a_{31}x_1 + a_{32}x_2 + a_{33}x_3 &= a_{34} \end{aligned} \tag{1}$$

with determinant

$$\Delta = \begin{vmatrix} a_{11} & a_{12} & a_{13} \\ a_{21} & a_{22} & a_{23} \\ a_{31} & a_{32} & a_{33} \end{vmatrix} \neq 0 \tag{2}$$

has the solution

$$x_1 = \frac{\begin{vmatrix} a_{14} & a_{12} & a_{13} \\ a_{24} & a_{22} & a_{23} \\ a_{34} & a_{32} & a_{33} \end{vmatrix}}{\Delta} \qquad x_2 = \frac{\begin{vmatrix} a_{11} & a_{14} & a_{13} \\ a_{21} & a_{24} & a_{23} \\ a_{31} & a_{34} & a_{33} \end{vmatrix}}{\Delta} \qquad x_3 = \frac{\begin{vmatrix} a_{11} & a_{12} & a_{14} \\ a_{21} & a_{22} & a_{24} \\ a_{31} & a_{32} & a_{34} \end{vmatrix}}{\Delta} \tag{3}$$

known as Cramer's rule.

When $a_{14} = a_{24} = a_{34} = 0$, (3) reduces to the system of linear homogeneous equations

$$\begin{aligned} a_{11}x_1 + a_{12}x_2 + a_{13}x_3 &= 0 \\ a_{21}x_1 + a_{22}x_2 + a_{23}x_3 &= 0 \\ a_{31}x_1 + a_{32}x_2 + a_{33}x_3 &= 0 \end{aligned} \tag{4}$$

and (3) yields solutions of the form

$$x_i = \frac{0}{\Delta} = 0 \qquad i = 1, 2, 3$$

For nontrivial solutions to exist, it is necessary, therefore, that $\Delta = 0$.

If not all coefficients a_{ij} in (4) are zero, two possibilities may then arise: (1) a second order determinant of Δ is not equal to zero; (2) all second order determinants are zero.

Suppose the second order determinant that is not equal to zero is

$$\Delta_2 = \begin{vmatrix} a_{11} & a_{12} \\ a_{21} & a_{22} \end{vmatrix} \neq 0 \tag{5}$$

Then the nonhomogeneous system in the two unknowns x_1 and x_2,

$$\begin{aligned} a_{11}x_1 + a_{12}x_2 &= -a_{13}x_3 \\ a_{21}x_1 + a_{22}x_2 &= -a_{23}x_3 \end{aligned} \tag{6}$$

may be solved for x_1, x_2 in terms of x_3. By Cramer's rule the solution is

$$x_1 = \frac{\begin{vmatrix} a_{12} & a_{13} \\ a_{22} & a_{23} \end{vmatrix}}{\Delta_2} x_3 \qquad x_2 = \frac{-\begin{vmatrix} a_{11} & a_{13} \\ a_{21} & a_{23} \end{vmatrix}}{\Delta_2} x_3 \tag{7}$$

or equivalently,

$$x_1 : x_2 : x_3 = \begin{vmatrix} a_{12} & a_{13} \\ a_{22} & a_{23} \end{vmatrix} : -\begin{vmatrix} a_{11} & a_{13} \\ a_{21} & a_{23} \end{vmatrix} : \begin{vmatrix} a_{11} & a_{12} \\ a_{21} & a_{22} \end{vmatrix} \tag{8}$$

This solution also satisfies the third equation of (4), for expanding determinant Δ in terms of the elements of the third row,

$$\Delta = a_{31}\begin{vmatrix} a_{12} & a_{13} \\ a_{22} & a_{23} \end{vmatrix} - a_{32}\begin{vmatrix} a_{11} & a_{13} \\ a_{21} & a_{23} \end{vmatrix} + \begin{vmatrix} a_{33} & a_{11} \\ a_{21} & a_{22} \end{vmatrix} a_{12} = 0$$

From (7) it is seen that to each arbitrary value of x_3 there corresponds a solution of (4). Thus, when $\Delta = 0$ and *at least* one second order determinant of Δ is not equal to zero, (4) has a *one* parameter family of solutions, called a general solution.

The possibility is considered now that all second order determinants of Δ are zero. Then, if at least one first order determinant of Δ, say a_{11}, equals zero, each equation of (4) is a multiple of the other and a solution of any one of the equations, say the first,

$$x_1 = -\frac{1}{a_{11}}(a_{12}x_2 + a_{13}x_3)$$

satisfies all three equations. Since arbitrary values may be given to x_2 and x_3, this solution depends on the two parameters x_2 and x_3.

The results just established are stated formally in the next theorem.

Theorem 8.2 **a.** *A necessary and sufficient condition that the system (4) of three linear homogeneous equations in three unknowns have solutions other than* 0, 0, 0 *is that the determinant*

$$\Delta = \begin{vmatrix} a_{11} & a_{12} & a_{13} \\ a_{21} & a_{22} & a_{23} \\ a_{31} & a_{32} & a_{33} \end{vmatrix}$$

of the system equals zero.

b. *If $\Delta = 0$ and a second order determinant of Δ is not equal to zero, the system has nontrivial solutions depending on a* single *arbitrary parameter.*

c. *If $\Delta = 0$ and all second order determinants of Δ equal zero, and if at least one of its first order determinants is not equal to zero, the system has nontrivial solutions depending on* two *arbitrary parameters.*

Corollary *A system of two linear homogeneous equations in three unknowns always has nontrivial solutions.*

The corollary is illustrated by solutions (7) of system (6).

The algebraic theory just presented is particularly useful in finding the point of meeting of projective lines corresponding to parallel lines in the Euclidean plane. For example, the system

$$\begin{aligned} -x_1 + x_2 + 2x_3 &= 0 \\ -2x_1 + 2x_2 + 5x_3 &= 0 \\ 3x_1 - 3x_2 + 7x_3 &= 0 \end{aligned}$$

with determinant

$$\Delta = \begin{vmatrix} -1 & 1 & 2 \\ -2 & 2 & 5 \\ 3 & -3 & 7 \end{vmatrix} = 0$$

has the solutions

$$x_1 : x_2 : x_3 = \begin{vmatrix} 1 & 2 \\ 2 & 5 \end{vmatrix} : -\begin{vmatrix} -1 & 2 \\ -2 & 5 \end{vmatrix} : \begin{vmatrix} -1 & 1 \\ -2 & 2 \end{vmatrix} = 1 : 1 : 0$$

Lines represented by the system are shown in Fig. 8.5. They meet at the ideal point (1,1,0) of the projective plane. In the Euclidean plane they are parallel lines and hence do not meet.

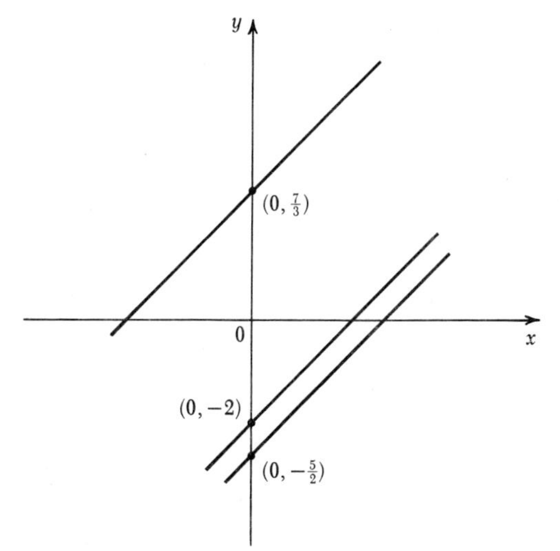

FIG. 8.5

Exercise 8.4

1 a. Show that the following system has nontrivial solutions:

$$\begin{aligned} 5x_1 + 14x_2 + 4x_3 &= 0 \\ 2x_1 - x_2 + x_3 &= 0 \\ 3x_1 + 4x_2 + 2x_3 &= 0 \end{aligned}$$

b. Find all solutions of the system.

2 a. Find a particular solution of the system

$$\begin{aligned} 2x_1 + 3x_2 + 6x_3 &= 0 \\ 3x_1 - 6x_2 + 2x_3 &= 0 \end{aligned}$$

b. On how many parameters does the general solution depend? Why?

c. Plot the lines represented by the system and give two sets of coordinates for their point of meeting.

3 Plot the lines

$$\begin{aligned} -x_1 + x_2 + 2x_3 &= 0 \\ -2x_1 + 2x_2 + 5x_3 &= 0 \\ 3x_1 - 3x_2 + 7x_3 &= 0 \end{aligned}$$

in an x, y coordinate system and find a set of homogeneous coordinates for their point of intersection.

8.7 Analytic expressions for collinearity of points and concurrency of lines

The algebraic results just obtained are carried over to geometry. Consider the line $[d_1,d_2,d_3]$ whose point equation is

$$d_1x_1 + d_2x_2 + d_3x_3 = 0 \tag{1}$$

If $P_1(a_{11},a_{12},a_{13})$, $P_2(a_{21},a_{22},a_{23})$, $P_3(a_{31},a_{32},a_{33})$ are three points of this line, substitution of the coordinates of the points for x_1, x_2, x_3 yields the system of equations in the three unknowns d_1, d_2, d_3, that is,

$$\begin{aligned} a_{11}d_1 + a_{12}d_2 + a_{13}d_3 &= 0 \\ a_{21}d_1 + a_{22}d_2 + a_{23}d_3 &= 0 \\ a_{31}d_1 + a_{32}d_2 + a_{33}d_3 &= 0 \end{aligned} \tag{2}$$

By Theorem 8.1 the system has nontrivial solutions if and only if

$$\Delta = \begin{vmatrix} a_{11} & a_{12} & a_{13} \\ a_{21} & a_{22} & a_{23} \\ a_{31} & a_{32} & a_{33} \end{vmatrix} = 0 \tag{3}$$

The vanishing of this determinant is, therefore, the necessary and sufficient condition that the three points are on a line.

Dually, (3) is also the necessary and sufficient condition for the concurrency of the three lines: $[a_{11},a_{12},a_{13}]$, $[a_{21},a_{22},a_{23}]$, $[a_{31},a_{32},a_{33}]$. Also

$$\begin{vmatrix} x_1 & x_2 & x_3 \\ a_{21} & a_{22} & a_{23} \\ a_{31} & a_{32} & a_{33} \end{vmatrix} = 0 \tag{4}$$

is the equation of the line through the two points (a_{21},a_{22},a_{23}) and (a_{31},a_{32},a_{33}), whereas

$$\begin{vmatrix} u_1 & u_2 & u_3 \\ a_{21} & a_{22} & a_{23} \\ a_{31} & a_{32} & a_{33} \end{vmatrix} = 0 \tag{5}$$

is the equation of a point on the two lines $[a_{21},a_{22},a_{23}]$, $[a_{31},a_{32},a_{33}]$.

Some examples will illustrate the theory.

Example 1 The three points $O(0,0,1)$, $I_x(0,1,0)$, $I_y(1,0,0)$ of Fig. 8.2 are not collinear since

$$\begin{vmatrix} 0 & 0 & 1 \\ 0 & 1 & 0 \\ 1 & 0 & 0 \end{vmatrix} \neq 0$$

Example 2 The three points $A(1,0,-1)$, $B(2,3,1)$, $C(-1,1,2)$ of Fig. 8.6 are collinear since

$$\begin{vmatrix} 1 & 0 & -1 \\ 2 & 3 & 1 \\ -1 & 1 & 2 \end{vmatrix} = 0$$

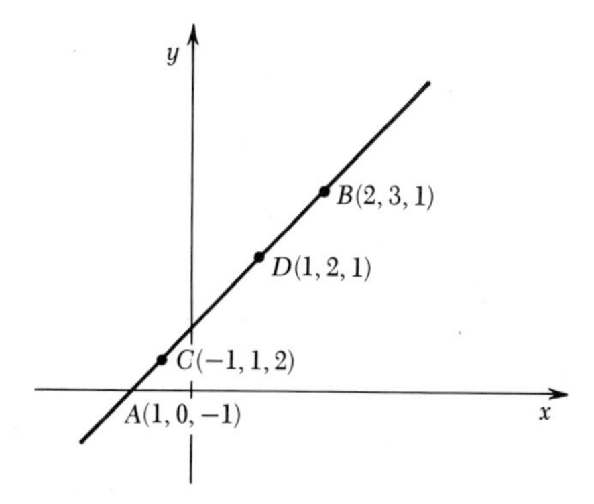

FIG. 8.6

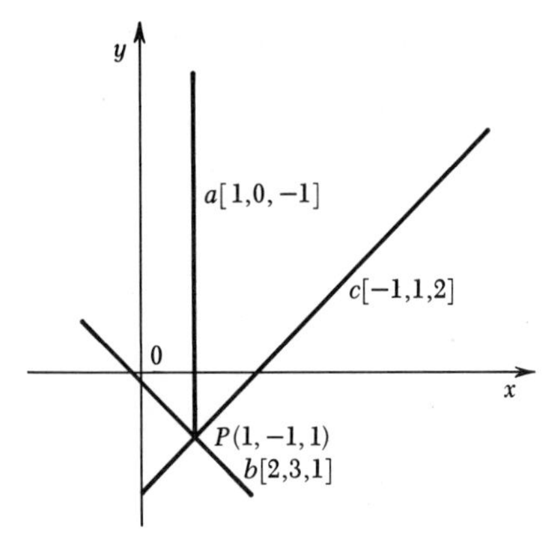

FIG. 8.7

and dually the three lines $a[1,0,-1]$, $b[2,3,1]$, $c[-1,1,2]$ (Fig. 8.7) are concurrent.

Example 3

$$\begin{vmatrix} x_1 & x_2 & x_3 \\ 1 & 0 & 0 \\ 0 & 0 & 1 \end{vmatrix} = 0 \qquad \text{or} \qquad x_2 = 0$$

is the equation of the line through the two points (1,0,0) and (0,0,1).

Example 4

$$\begin{vmatrix} u_1 & u_2 & u_3 \\ 1 & 0 & 0 \\ 0 & 0 & 1 \end{vmatrix} = 0 \qquad \text{or} \qquad u_2 = 0$$

is the equation of the point on the two lines [1,0,0] and [0,0,1].

Exercise 8.5

1 Are the points

$$A(2,3,1) \qquad B(5,-2,2) \qquad C(1,-8,0)$$

collinear? Why?

2 For what value of k are the lines

$$\begin{aligned} x_1 - x_2 + x_3 &= 0 \\ kx_1 + x_2 + 3x_3 &= 0 \\ x_1 - x_2 + 3x_3 &= 0 \end{aligned}$$

concurrent? Explain.

3 Write in determinant form (a) the point equation of the line through points $P_1(a_{11},a_{12},a_{13})$, $P_2(a_{21},a_{22},a_{23})$ and (b) the line equation of the point on the lines $l_1[a_{11},a_{12},a_{13}]$, $l_2[a_{21},a_{22},a_{23}]$.

8.8 Matrices

For generalizing the algebraic theory of Sec. 8.6, matrices will be found useful.

By definition, a rectangular array A of real numbers consisting of m rows and n columns of the form

$$A = \begin{pmatrix} a_{11} & a_{12} & \cdots & a_{1n} \\ a_{21} & a_{22} & \cdots & a_{2n} \\ \cdots & \cdots & \cdots & \cdots \\ a_{m1} & a_{m2} & \cdots & a_{mn} \end{pmatrix} = a_{ij} \qquad \begin{matrix} i = 1, \ldots, m \\ j = 1, \ldots, n \end{matrix}$$

is an $m \times n$ matrix over the field of real numbers. The symbol a_{ij}, where $i = 1, \ldots, m$ and $j = 1, \ldots, n$, is also used to denote the matrix A.

A matrix, every element of which is zero, is called the zero *matrix.*

Examples of matrices are the respective column matrix B, the row matrix C, the 2×3 matrix D, and the 2×2 zero matrix E.

$$B = \begin{pmatrix} x_1 \\ x_2 \end{pmatrix} \qquad C = (y_1 y_2) \qquad D = \begin{pmatrix} 1 & 2 & 1 \\ 0 & 7 & 5 \end{pmatrix} \qquad E \begin{pmatrix} 0 & 0 \\ 0 & 0 \end{pmatrix}$$

Two $m \times n$ matrices $A = (a_{ij})$ and $B = (b_{ij})$ are equal if and only if

$$a_{ij} = b_{ij}$$

for every i and every j.

For example, the two matrices

$$D \begin{pmatrix} 1 & 2 & 1 \\ 0 & 7 & 5 \end{pmatrix} \qquad \text{and} \qquad F = \begin{pmatrix} 1 & 2 & k \\ 0 & l & 5 \end{pmatrix}$$

are equal if and only if $k = 1$ and $l = 7$.

If $m = n$, matrix A is square and its determinant is

$$\Delta = \begin{vmatrix} a_{11} & a_{12} & \cdots & a_{1n} \\ a_{21} & a_{22} & \cdots & a_{2n} \\ \cdots & \cdots & \cdots & \cdots \\ a_{n1} & a_{n2} & \cdots & a_{nn} \end{vmatrix}$$

For reasons to be explained later, the square $n \times n$ matrix I

$$I = \begin{pmatrix} 1 & 0 & \cdots & 0 \\ 0 & 1 & \cdots & 0 \\ \cdots & \cdots & \cdots & \cdots \\ 0 & 0 & \cdots & 1 \end{pmatrix}$$

is called the identity matrix. *A square matrix whose determinant is not equal to zero is said to be nonsingular, and otherwise singular.*

8.9 Rank of a matrix

A matrix A even though square is not a determinant. It is, to repeat, simply an array of numbers. However, if $n, m > 1, 1$, numerous determinants may be formed from A by suppressing appropriate rows and/or columns.

For example, from the matrix B

$$B = \begin{pmatrix} a_{11} & a_{21} & a_{31} \\ a_{12} & a_{22} & a_{32} \end{pmatrix} \tag{1}$$

may be formed the three second order determinants

$$\begin{vmatrix} a_{11} & a_{21} \\ a_{12} & a_{22} \end{vmatrix} \qquad \begin{vmatrix} a_{11} & a_{31} \\ a_{12} & a_{32} \end{vmatrix} \qquad \begin{vmatrix} a_{21} & a_{31} \\ a_{22} & a_{32} \end{vmatrix} \tag{2}$$

and the rank of a matrix is introduced with the definition:

Definition *If, of all the determinants that can be formed from a given matrix A, not all those of order r are zero but all those of order $s > r$ are zero, the rank of A is r.* (Zero rank means that all elements of the matrix are zero.)

For example, the rank of the matrix C

$$C = \begin{pmatrix} 2 & 4 & 6 \\ 1 & 2 & k \end{pmatrix} \tag{3}$$

is 1 if $k = 3$ and is 2 if $k \neq 3$.

Also, the matrix

$$\begin{pmatrix} -1 & 1 & 2 \\ -2 & 2 & 5 \\ 3 & -3 & 7 \end{pmatrix} \tag{4}$$

with determinant $\Delta = 0$ has rank $r = 2$ since

$$\begin{vmatrix} 1 & 2 \\ 2 & 5 \end{vmatrix} \neq 0 \tag{5}$$

A reference to the system of m linear homogeneous equations in the n unknowns $x_1, x_2, \ldots, x_n$,

$$\begin{aligned} a_{11}x_1 + a_{12}x_2 + \cdots + a_{1n}x_n &= 0 \\ a_{21}x_1 + a_{22}x_2 + \cdots + a_{2n}x_n &= 0 \\ \cdots\cdots\cdots\cdots\cdots\cdots\cdots\cdots \\ a_{m1}x_1 + a_{m2}x_2 + \cdots + a_{mn}x_n &= 0 \end{aligned} \tag{6}$$

shows that matrix A (Sec. 8.8) is simply the array of coefficients in these equations. When such is the case, A is called the *matrix of system.*

By definition the system (6) *has rank r if the rank of A is r.* The definition permits the determinant tests of Theorem 8.1 to be put in the more elegant and compact form:

Theorem 8.3
a. *A necessary and sufficient condition that system* (4) *of Sec.* 8.6 *have solutions other than* 0, 0, 0 *is that rank* $r < 3$.
b. *If* $r = 2$, *the system has nontrivial solutions depending on a single arbitrary parameter.*
c. *If* $r = 1$, *the system has nontrivial solutions depending on two arbitrary parameters.*

An extension of the arguments of Sec. 8.6 to system (6) gives the results:

Theorem 8.4
a. *A system of m linear homogeneous equations in n unknowns* [*system* (6)] *has nontrivial solutions if* $m < n$.
b. *If* $m \geq n$, *the system* (6) *has nontrivial solutions if an only if the rank r of the system* $< n$.

Example 1 The system of m = two equations in n = three unknowns

$$\begin{aligned} 2x_1 + 0x_2 + x_3 &= 0 \\ 3x_1 + x_2 + x_3 &= 0 \end{aligned}$$

with matrix

$$\begin{pmatrix} 2 & 0 & 1 \\ 3 & 1 & 1 \end{pmatrix}$$

of rank $r = 2$ *has nontrivial* solutions.

Example 2 The system of m = three equations in n = two unknowns

$$\begin{aligned} 2x_1 + 3x_2 &= 0 \\ 0x_1 + x_2 &= 0 \\ x_1 + x_2 &= 0 \end{aligned}$$

with matrix

$$\begin{pmatrix} 2 & 3 \\ 0 & 1 \\ 1 & 1 \end{pmatrix}$$

of rank $r = 2$ has *only the trivial solution* (0,0).

8.10 Operations on matrices

For later use it will be necessary to define two operations on matrices: (1) multiplication of a matrix by a scalar, called scalar multiplication, and (2) multiplication of a matrix by another matrix, called matrix multiplication.

Definition 1 *Multiplication of a matrix A by a scalar c is the matrix cA in which every element of A is multiplied by c.*

For example, if A and B are the respective 2×1 and 2×3 matrices

$$A = \begin{pmatrix} a_{11} \\ a_{21} \end{pmatrix} \qquad \text{and} \qquad B = \begin{pmatrix} a_{11} & a_{12} & a_{13} \\ a_{21} & a_{22} & a_{23} \end{pmatrix}$$

then

$$CA = \begin{pmatrix} ca_{11} \\ ca_{21} \end{pmatrix} \qquad \text{and} \qquad CB = \begin{pmatrix} ca_{11} & ca_{12} & ca_{13} \\ ca_{21} & ca_{22} & ca_{23} \end{pmatrix}$$

Definition 2 *If $A = a_{ij}$ is the $m \times n$ matrix and $B = b_{jk}$ the $n \times r$ matrix, the product AB is the $m \times r$ matrix $C = c_{ik}$, where*

$$c_{ik} = a_{i1}b_{ik} + a_{i2}b_{2k} + a_{in}b_{nk} = \sum_{j=1}^{n} a_{ij}b_{jk}$$

From definition 2 it is seen that matrix multiplication AB is possible only *if the number of columns* of A *equals the number of rows* of B. Also, matrix multiplication does not in general obey the commutative law $AB = BA$.

Some examples will illustrate these statements. If

$$A = \begin{pmatrix} a_{11} & a_{12} \\ a_{21} & a_{22} \end{pmatrix} \qquad B = \begin{pmatrix} x_1 \\ x_2 \end{pmatrix} \qquad C = \begin{pmatrix} c_{11} & c_{12} & c_{13} \\ c_{21} & c_{22} & c_{23} \\ c_{31} & c_{32} & c_{33} \end{pmatrix}$$

AB is the 2×1 matrix

$$\begin{pmatrix} a_{11}x_1 + a_{12}x_2 \\ a_{21}x_1 + a_{22}x_2 \end{pmatrix}$$

which incidentally should not be confused with the 2×2 matrix

$$\begin{pmatrix} a_{11}x_1 & a_{12}x_2 \\ a_{21}x_1 & a_{22}x_2 \end{pmatrix}$$

The product AC is not possible since A has two columns and C has three rows.

To illustrate noncommutativity, let

$$D = \begin{vmatrix} 1 & 0 \\ 2 & -1 \end{vmatrix} \qquad \text{and} \qquad E = \begin{vmatrix} 1 & -1 \\ 0 & 1 \end{vmatrix}$$

Then

$$DE = \begin{vmatrix} 1 & -1 \\ 0 & -1 \end{vmatrix} \qquad ED = \begin{vmatrix} 1 & -1 \\ 0 & 1 \end{vmatrix} \qquad \text{and} \qquad DE \neq ED$$

Now that products of matrices have been defined, system (1) of Sec. 8.6 may be written in the more compact matrix form

$$A \cdot X = B$$

where $A = \begin{pmatrix} a_{11} & a_{12} & a_{13} \\ a_{21} & a_{22} & a_{23} \\ a_{31} & a_{32} & a_{33} \end{pmatrix}$

$$X = \begin{pmatrix} x_1 \\ x_2 \\ x_3 \end{pmatrix}$$

$$B = \begin{pmatrix} a_{14} \\ a_{24} \\ a_{34} \end{pmatrix}$$

It is left as an exercise for the reader to show that matrix multiplication obeys the associative axiom $A(BC) = (AB)C$.

8.11 The identity and inverse matrices

The particular square matrix

$$I = \begin{pmatrix} 1 & 0 & \cdots & 0 \\ 0 & 1 & \cdots & 0 \\ \cdots & \cdots & \cdots & \cdots \\ 0 & 0 & \cdots & 1 \end{pmatrix} \tag{1}$$

has already been called the identity matrix. There is a reason for the name. If A is the simple 2×2 matrix

$$A = \begin{pmatrix} a_{11} & a_{12} \\ a_{21} & a_{22} \end{pmatrix}$$

the products

$$\begin{pmatrix} 1 & 0 \\ 0 & 1 \end{pmatrix}\begin{pmatrix} a_{11} & a_{12} \\ a_{21} & a_{22} \end{pmatrix} = \begin{pmatrix} a_{11} & a_{12} \\ a_{21} & a_{22} \end{pmatrix}$$

and

$$\begin{pmatrix} a_{11} & a_{12} \\ a_{21} & a_{22} \end{pmatrix}\begin{pmatrix} 1 & 0 \\ 0 & 1 \end{pmatrix} = \begin{pmatrix} a_{11} & a_{12} \\ a_{21} & a_{22} \end{pmatrix}$$

may be written

$$IA = A = AI \tag{2}$$

a relation which holds also if A is an $n \times n$ matrix, $n > 0$. The relation shows that matrix I has the property of the number 1 in ordinary multiplication, where if x is any real number

$$1 \cdot x = x = x \cdot 1$$

In other words, the result of multiplication of matrix A by matrix I, or I by A, is again the matrix A. Because of this property, I is called the *identity matrix*. It is used now to define an extremely important matrix:

Definition *If A is a square matrix for which there exists a matrix A^{-1} such that $A \cdot A^{-1} = A^{-1} \cdot A = I$, A^{-1} is called an inverse of the matrix A.*

To form, for example, the inverse A^{-1} of a simple matrix, say

$$\mathrm{A} = \begin{pmatrix} a_{11} & a_{12} \\ a_{21} & a_{22} \end{pmatrix}$$

let

$$\mathrm{A}^{-1} = \begin{pmatrix} b_{11} & b_{12} \\ b_{21} & b_{22} \end{pmatrix}$$

Then

$$\mathrm{A} \cdot \mathrm{A}^{-1} = \begin{pmatrix} a_{11}b_{11} + a_{12}b_{21} & a_{11}b_{12} + a_{12}b_{22} \\ a_{21}b_{11} + a_{22}b_{21} & a_{21}b_{12} + a_{22}b_{22} \end{pmatrix} = \begin{pmatrix} 1 & 0 \\ 0 & 1 \end{pmatrix}$$

and hence elements of A^{-1} must satisfy the system of equations

$$\begin{aligned} a_{11}b_{11} + a_{12}b_{21} &= 1 \\ a_{11}b_{12} + a_{12}b_{22} &= 0 \\ a_{21}b_{11} + a_{22}b_{21} &= 0 \\ a_{21}b_{12} + a_{22}b_{22} &= 1 \end{aligned}$$

For the particular matrix

$$\mathrm{A} = \begin{pmatrix} 2 & 3 \\ 1 & -1 \end{pmatrix}$$

this system reduces to

$$\begin{aligned} 2b_{11} + 3b_{21} &= 1 \\ b_{11} - b_{21} &= 0 \\ 2b_{12} + 3b_{22} &= 0 \\ b_{12} - b_{22} &= 1 \end{aligned}$$

with solutions

$$b_{11} = b_{21} = 1/5 \qquad b_{12} = 3/5 \qquad b_{22} = -2/5$$

Hence

$$\mathrm{A}^{-1} = \begin{pmatrix} 1/5 & 3/5 \\ 1/5 & -2/5 \end{pmatrix}$$

is the inverse of

$$A = \begin{pmatrix} 2 & 3 \\ 1 & -1 \end{pmatrix}$$

Note how laborious is the process just described. Fortunately there are powerful machines which, when properly operated, grind out the inverse mechanically. The mathematician and the inventor have again joined forces in the interest of science.

The inverse matrix is put to good use in solving the system of equations

$$\begin{aligned} a_{11}x_1 + a_{12}x_2 &= kx_1' \\ a_{21}x_1 + a_{22}x_2 &= kx_2' \end{aligned} \qquad k \neq 0 \tag{3}$$

for x_1, x_2 in terms of x_1', x_2'.

If

$$A = \begin{pmatrix} a_{11} & a_{12} \\ a_{21} & a_{22} \end{pmatrix} \qquad X = \begin{pmatrix} x_1 \\ x_2 \end{pmatrix} \qquad X' = \begin{pmatrix} x_1' \\ x_2' \end{pmatrix}$$

system (3) in matrix form is

$$AX = kX' \tag{4}$$

and multiplication by A^{-1} yields the matrix equation

$$A^{-1}(AX) = A^{-1}(kX') \tag{5}$$

From the associative axiom and the definition of an inverse, (5) reduces to the matrix solution

$$X = k(A^{-1})X' \tag{6}$$

For example, the particular system

$$\begin{aligned} 2x_1 + 3x_2 &= kx_1' \\ x_1 - x_2 &= kx_2' \end{aligned} \tag{7}$$

has the matrix solution

$$X = kA^{-1}X' \qquad A^{-1} = \begin{pmatrix} 1/5 & 3/5 \\ 1/5 & -2/5 \end{pmatrix}$$

which, in more familiar form is

$$\begin{aligned} x_1 &= k(1/5x_1' + 3/5x_2') \\ x_2 &= k(1/5x_1' - 2/5x_2') \end{aligned} \tag{8}$$

Exercise 8.6

1 If

$$A = \begin{pmatrix} 2 & 3 \\ 3 & 5 \end{pmatrix} \qquad B = \begin{pmatrix} 4 & 2 \\ 2 & 1 \end{pmatrix}$$

find AB and BA. Are A and B inverse matrices? Why?

2 Are there matrices other than inverse matrices which satisfy the commutative law $AB = BA$? Explain.

3 If $E = (21)$ and $F = (111)$ are two row matrices, is the product EF possible? Why?

4 If AB is the zero matrix, is either A or B the zero matrix? Explain.

5 Show that the inverse of

$$\begin{pmatrix} 1 & 2 & 3 \\ 1 & 3 & 3 \\ 1 & 2 & 4 \end{pmatrix}$$

is

$$\begin{pmatrix} 6 & -2 & -3 \\ -1 & 1 & 0 \\ -1 & 0 & 1 \end{pmatrix}$$

6 Write in matrix form the system of equations

$$\begin{aligned} -x_1 + x_2 + 2x_3 &= x_1' \\ -2x_1 + 2x_2 + 5x_3 &= x_2' \\ 3x_1 - 3x_2 + 7x_3 &= x_3' \end{aligned}$$

and using the inverse of the matrix of the system, express x_1, x_2, x_3 in terms of x_1', x_2', x_3'.

7 Prove that if A is a nonsingular matrix and if $B = AC$, then B and C have the same rank. *Hint:* See ([50] p. 98).

8 Prove that all nonsingular $n \times n$ matrices form a group with respect to the operation of multiplication of matrices.

Concluding remarks

The material of this chapter is a first extension of Descartes' coordinate geometry of the Euclidean plane. From the rectangular nonhomogeneous coordinates x, y of a point P of the Euclidean plane, one passes to the homogeneous coordinates x_1, x_2, x_3 of a point of the projective plane by the substitution

$$x = \frac{x_1}{x_3} \qquad y = \frac{x_2}{x_3} \qquad x_3 \neq 0$$

and points for which $x_3 = 0$ are called *ideal* points.

With the introduction of homogeneous line coordinates, duality in algebraic geometry is exhibited by means of the equation

$$u_1x_1 + u_2x_2 + u_3x_3 = 0$$

which represents the point equation of a line if the u_i's are fixed and the x_i's are variables, or the line equation of a point if the x_i's are fixed and the u_i's are variables.

Systems of m linear equations in n unknowns, $m \gtreqless n$, were then studied and their solutions used in determining, or setting up, simple algebraic tests for collinearity of points and concurrency of lines. Seeing algebraic and geometric theory thus joined makes both studies more enjoyable.

Concluding sections were then devoted to a study of matrices and matrix theory showing how the new concepts are used to simplify, extend, and round out the algebraic theory. Although the discussion is necessarily brief, it does give some indication of how matrix theory can be used to eliminate many of the cumbersome, awkward algebraic processes of former years. As for technical facility in matrix operation, it can be acquired in the usual fashion by practice.[1]

All the material is preparation for still more interesting things to come. The inexorable force that pushed out the early frontiers of Descartes' coordinate geometry continues to extend them.

Suggestions for further reading

Finkbeiner, David T., II: "Introduction to Matrices and Linear Transformations," chaps. 2 to 4.

[1] For an elementary approach to matrices see the recent publication "The Mathematics of Matrices" by Philip J. Davis, Blaisdell Publishing Co., New York, 1965.

Fishbach, W. T.: "Projective and Euclidean Geometry," chaps. 6 and 7.
Graustein, W. E.: "Introduction to Higher Geometry," chaps. 3 and 5.
Hausner, Melvin: "A Vector Space Approach to Geometry," chap. 10.
Robinson G. de B.: "Vector Geometry," chap. 3.
Schreier D., and E. Sperner: "Introduction to Modern Algebra and Matrix Theory."
Smith, P. A., and A. S. Gale: "New Analytic Geometry."

9 vector theory in projective analytic geometry

The classical approach to coordinate geometry given in Chap. 8 is a natural transition from coordinate geometry of the Euclidean plane to that of the projective plane, but it leaves much to be desired. Among other things, the metric concepts used there are not in keeping with the present nonmetric development. The modern vector approach with its accompanying matrix theory is a decided improvement. Simplification, clarification, and generalization are a few of its many advantages, as the following material will show.

Much of what is presented in this and the next chapter will be found useful in areas other than the geometry for which it is intended. Linear algebra is one of them. A geometric background illuminates this all-important new development, and classical geometry is in turn enriched, clarified, and generalized by the modern theory.

Basic to all that follows are the notions of a vector, a vector space, and linear dependence of vectors. These concepts are discussed first.

9.1 Vectors

Originally a vector was defined as a quantity having both magnitude and direction, and vectors were used in the treatment of such physical concepts as force, velocity, and acceleration.

In Fig. 9.1 a vector is represented by the directed line segment **OP** joining the origin O of a nonhomogeneous coordinate system to a point $P(x,y,z)$ in space. Since magnitude and direction are determined once the coordinates x, y, z of P are given, the vector is determined equally well by the triple of numbers x, y, z, and there is thus illustrated the modern definition of a vector:

Definition 1 *A set of n real numbers $x_1, x_2, \ldots, x_n$ will be called a vector in the space of n-tuples, and $x_1, x_2, \ldots, x_n$ are its components.*

The number pair (0,1) represents a vector in the space of 2-tuples, and the number triple $(-1,2,-7)$ represents a vector in the space of 3-tuples.

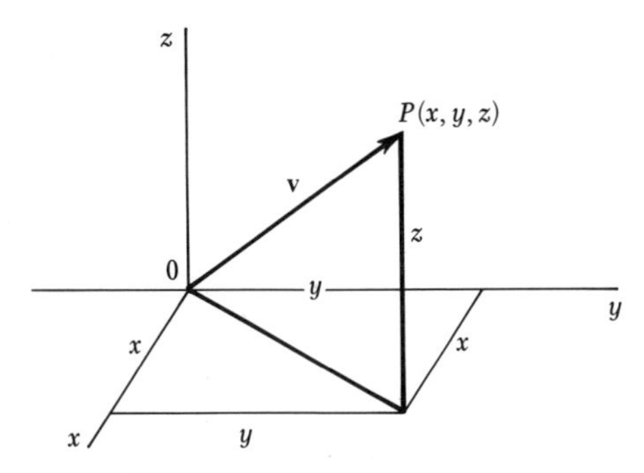

FIG. 9.1

9.2 Vector space

A mathematical system will be described now in which S is the set of all possible n-tuples of real numbers representing vectors

$$\mathbf{v}_1(x_{11},x_{12}, \ldots ,x_n),\ \mathbf{v}_2(x_{21},x_{22}, \ldots ,x_{2n}), \ldots ,\ \mathbf{v}_n(x_{n1},x_{n2}, \ldots ,x_{nn})$$

A boldface letter is used to distinguish a vector from a scalar.

To introduce an algebra into the set S, let the zero vector be the one with all of its components zero. It is denoted simply by $\theta(0,0, \ldots ,0)$. The following vectors $\epsilon_1(1,0, \ldots ,0)$, $\epsilon_2(0,1, \ldots ,0)$, . . . , $\epsilon_n(0,0, \ldots , 1)$ are called natural vectors.

Two vectors

$$\mathbf{A}(a_1,a_2, \ldots ,a_n) \qquad \text{and} \qquad \mathbf{B}(b_1,b_2, \ldots ,b_n)$$

are said to be *equal if and only if* $a_i = b_i$, $i = 1, 2, \ldots , n$.

Two operations on elements of S are vector *addition* and scalar *multiplication* indicated by the respective symbols $+$ and $\cdot$ and defined as follows:

$$\mathbf{v}_1(x_1,x_2, \ldots ,x_n) + \mathbf{v}_2(y_1,y_2, \ldots ,y_n) = \mathbf{v}_3(x_1 + y_1,\ x_2 + y_2, \ldots ,x_n + y_n)$$

and

$$c\mathbf{v}_1 = \mathbf{v}_y(cx_1,cx_2, \ldots ,cx_n)$$

where c is any real number.

For example,

$$\mathbf{v}_1(2,0) + \mathbf{v}_2(-1,2) = \mathbf{v}_3(1,2) \qquad \text{(Fig. 9.2)}$$
$$2\mathbf{v}_1(2,1) = \mathbf{v}_2(4,2) \qquad \text{(Fig. 9.3)}$$

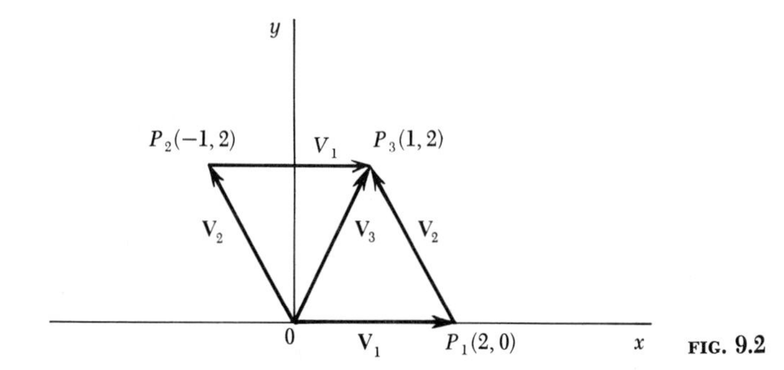

FIG. 9.2

The assumption is made now that the operation of vector addition satisfies the axioms for a commutative group:

1	$\mathbf{v}_1 + \mathbf{v}_2 \in S$	(Closure)
2	$(\mathbf{v}_1 + \mathbf{v}_2) + \mathbf{v}_3 = \mathbf{v}_1 + (\mathbf{v}_2 + \mathbf{v}_3)$	(Associative law)
3	$\mathbf{v}_1 + \mathbf{v}_2 = \mathbf{v}_2 + \mathbf{v}_1$	(Commutative law)
4	$\mathbf{v}_1 + (-\mathbf{v}_1) = \theta$	($-\mathbf{v}_1$ is the inverse of $\mathbf{v}_1$ and θ is the identity element of S)

Also, if c_1, c_2, c_3 are real numbers, scalar multiplication satisfies the additional axioms:

5 $c(\mathbf{v}_1 + \mathbf{v}_2) = c\mathbf{v}_1 + c\mathbf{v}_2$
6 $(c_1 + c_2)\mathbf{v} = c_1\mathbf{v} + c_2\mathbf{v}$
7 $(c_1c_2)\mathbf{v} = c_1(c_2\mathbf{v})$

Definition 2 *A set of elements satisfying these axioms is called the vector space V of n-tuples.*

FIG. 9.3

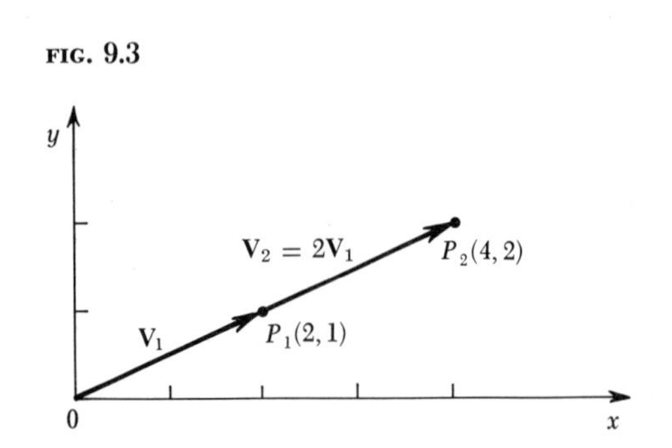

FIG. 9.4

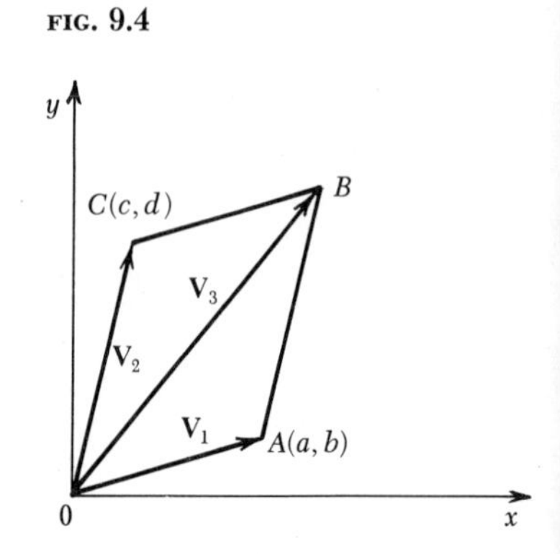

Definition 3 *If any subset of V forms a vector space W in itself, W is called a subspace of V.*

For example, all vectors $\mathbf{v}(x_1,x_2,x_3)$ in the space of number triples, with $x_3 = 0$, form a *subspace* of V. They are illustrated by vectors all lying in the x_1x_2 plane.

Exercise 9.1

1 If coordinates (a,b) and (c,d) of points A and C of (Fig. 9.4) are components of vectors $\mathbf{v}_1$ and $\mathbf{v}_2$, and if $\mathbf{v}_3$ is the vector represented by the directed line segment OB and satisfying the condition:

$$\mathbf{v}_3 = \mathbf{v}_1 + \mathbf{v}_2$$

a. Show that $\Delta OAB \cong \Delta OCB$, and

b. Give components of $\mathbf{v}_3$

2 Show, in an x, y, z rectangular coordinate system, each of the vectors $\mathbf{v}_1(1,1,1)$, $\mathbf{v}_2(-1,-1,-1)$, and $\mathbf{v}_3 = \mathbf{v}_1 + \mathbf{v}_2$.

3 a. Give components of the sum of the two vectors

$$\mathbf{v}_1(1,0,1,1) \qquad \text{and} \qquad \mathbf{v}_2(-1,0,1,-1)$$

b. If $\mathbf{v}_3$ is the vector $(-1,0,0,0)$, give components of the vectors

$$\mathbf{v}_4 = (\mathbf{v}_1 + \mathbf{v}_2) + \mathbf{v}_3 \qquad \text{and} \qquad \mathbf{v}_5 = (\mathbf{v}_1 + \mathbf{v}_3) + \mathbf{v}_2$$

4 If $\mathbf{v}_1(x,0,z) + \mathbf{v}_2(2,-y,3z) = \mathbf{v}_3(5,6,9)$, find x, y, z.

9.3 Linear dependence of vectors

A formal definition introduces *linear dependence,* a basic and highly useful relation among vectors.

The vectors $\mathbf{v}_1, \mathbf{v}_2, \ldots, \mathbf{v}_m$ *are linearly dependent if constants* $k_1, k_2, \ldots, k_m \neq 0, 0, \ldots, 0$ *exist such that*

$$k_1\mathbf{v}_1 + k_2\mathbf{v}_2 + \cdots + k_m\mathbf{v}_m = \theta = \sum_{i=1}^{m} k_i\mathbf{v}_i \qquad (1)$$

or equivalently, the vectors $\mathbf{v}_1, \mathbf{v}_2, \ldots, \mathbf{v}_m$ *are linearly independent if* (1) *implies* $k_1 = k_2, \ldots, k_m = 0$.

If $\mathbf{v}_i$ has the n components $(a_{i1}, a_{i2}, \ldots, a_{in})$, and $i = 2, \ldots, m$, the vector equation (1) yields the system of n linear homogeneous equations in m unknowns:

$$\begin{array}{l} a_{11}k_1 + a_{21}k_2 + \cdots + a_{m1}k_m = 0 \\ a_{12}k_1 + a_{22}k_2 + \cdots + a_{m2}k_m = 0 \\ \cdots\cdots\cdots\cdots\cdots\cdots\cdots\cdots \\ a_{1n}k_1 + a_{2n}k_2 + \cdots + a_{mn}k_m = 0 \end{array} \tag{2}$$

and the problem of determining the linear dependence of vectors reduces to that of determining whether or not this system has a solution other than the trivial one $k_1 = k_2, \ldots, k_m = 0$.

The algebraic theory concerning these solutions was given in the last chapter. Some examples will illustrate the theory.

Example 1 The vectors $\mathbf{v}_1(1,0,-1)$, $\mathbf{v}_2(2,3,1)$, $\mathbf{v}_3(-1,1,2)$ are linearly dependent since the vector equation $k_1\mathbf{v}_1 + k_2\mathbf{v}_2 + k_3\mathbf{v}_3 = \theta$ yields the system of equations

$$\begin{aligned} k_1 + 2k_2 - k_3 &= 0 \\ 0k_1 + 3k_2 + k_3 &= 0 \\ -k_1 + k_2 + 2k_3 &= 0 \end{aligned}$$

with the nontrivial solutions

$$k_1\!:\!k_2\!:\!k_3 = \begin{vmatrix} 2 & -1 \\ 3 & 1 \end{vmatrix} : -\begin{vmatrix} 1 & -1 \\ 0 & 1 \end{vmatrix} : \begin{vmatrix} 1 & 2 \\ 0 & 3 \end{vmatrix}$$

Example 2 The natural vectors

$$\epsilon_1(1,0,\ldots,0),\ \epsilon_2(0,1,\ldots,0),\ \ldots,\ \epsilon_n(0,0,\ldots,1)$$

are linearly *independent* since the vector equation

$$k_1\epsilon_1 + k_2\epsilon_2 + \cdots + k_n\epsilon_n = \theta$$

is satisfied only when

$$k_1 = k_2 = \cdots = k_n = 0$$

Example 3 The two vectors

$$\mathbf{v}_1(1,2,3) \qquad \text{and} \qquad \mathbf{v}_2(0,1,1)$$

are linearly *independent* since the vector equation

$$k_1\mathbf{v}_1 + k_2\mathbf{v}_2 = \theta$$

yields the system of *three* equations in *two* unknowns, that is,

$$\begin{aligned} k_1 + 0k_2 &= 0 \\ 2k_1 + k_2 &= 0 \\ 3k_1 + k_2 &= 0 \end{aligned}$$

The system has only the trivial solution

$$k_1 = k_2 = 0$$

Example 4 The three vectors

$$\mathbf{v}_1(2,3) \qquad \mathbf{v}_2(0,1) \qquad \mathbf{v}_3(1,1)$$

are linearly dependent since the vector equation

$$k_1\mathbf{v}_1 + k_2\mathbf{v}_2 + k_3\mathbf{v}_3 = \theta$$

yields the system of two equations in three unknowns, that is,

$$\begin{aligned} 2k_1 + 0k_2 + k_3 &= 0 \\ 3k_1 + k_2 + k_3 &= 0 \end{aligned}$$

with the nontrivial solutions

$$k_1 : k_2 : k_3 = \begin{vmatrix} 0 & 1 \\ 1 & 1 \end{vmatrix} : -\begin{vmatrix} 2 & 1 \\ 3 & 1 \end{vmatrix} : \begin{vmatrix} 2 & 0 \\ 3 & 1 \end{vmatrix}$$

Exercise 9.2

1 Show that the vectors

$$\mathbf{v}_1(2,3,-2) \qquad \mathbf{v}_2(4,5,2) \qquad \mathbf{v}_3(1,2,-4)$$

are linearly dependent by finding constants k_1, k_2, $k_3 \neq 0, 0, 0$ satisfying

$$k_1\mathbf{v}_1 + k_2\mathbf{v}_2 + k_3\mathbf{v}_3 = \theta$$

2 Test for linear dependence each set of vectors

a. $\mathbf{v}_1(0,2,3)$, $\mathbf{v}_2(1,0,1)$, $\mathbf{v}_3(2,3,4)$

b. $\mathbf{v}_1(1,1,1,1)$, $\mathbf{v}_2(0,1,0,2)$, $\mathbf{v}_3(1,-1,0,1)$

c. $\mathbf{v}_1(1,1,1)$, $\mathbf{v}_2(0,1,2)$, $\mathbf{v}_3(-1,1,1)$, $\mathbf{v}_4(2,0,1)$

3 Write a system of four linear homogeneous equations in three unknowns, with numerical coefficients and having nontrivial solutions. Find the solutions.

4 Under what condition does the following system of three linear equations in the four unknowns k_1, k_2, k_3, k_4 have nontrivial solutions?

$$\begin{aligned} a_{11}k_1 + a_{12}k_2 + a_{13}k_3 + a_{14}k_4 &= 0 \\ a_{21}k_1 + a_{22}k_2 + a_{23}k_3 + a_{24}k_4 &= 0 \\ a_{31}k_1 + a_{32}k_2 + a_{33}k_3 + a_{34}k_4 &= 0 \end{aligned}$$

5 Prove that any set of vectors containing the zero vector is a linearly dependent set.

6 If $c_1, c_2, \ldots, c_m$ are arbitrary scalars, prove that according as $\mathbf{v}_1, \mathbf{v}_2, \ldots, \mathbf{v}_m$ are linearly dependent or independent, so are $c_1\mathbf{v}_1, c_2\mathbf{v}_2, \ldots, c_m\mathbf{v}_m$.

9.4 Bases for vector spaces. Dimension of a space

In the vector space of n-tuples, there exists linearly independent sets of vectors. One such set is the set of natural vectors

$$\epsilon_1(1,0,\ldots,0),\ \epsilon_2(0,1,\ldots,0),\ \ldots,\ \epsilon_n(0,0,\ldots,1)$$

in terms of which any vector in the space of n-tuples may be expressed thus:

$$\mathbf{v}(x_1,x_2,\ldots,x_n) = x_1\epsilon_1 + x_2\epsilon_2 + \cdots + x_n\epsilon_n$$

If now $\mathbf{v}_1, \mathbf{v}_2, \ldots, \mathbf{v}_n$ is another linearly independent set of vectors, from Theorem 8.4 this new set and any other vector $\mathbf{v}$ of the space form a linearly dependent set. Constants $k_1, k_2, \ldots, k_n \neq 0, 0, \ldots, 0$ therefore exist such that

$$\sum_{i=1}^{n} k_i\mathbf{v}_i + k_{n+1}\mathbf{v} = \theta$$

Now $k_{n+1} \neq 0$ (Why?), and hence $\mathbf{v}$ may be expressed thus:

$$\mathbf{v} = c_1\mathbf{v}_1 + c_2\mathbf{v}_2 + \cdots + c_n\mathbf{v}_n = \sum_{i=1}^{n} c_i\mathbf{v}_i \qquad c_i = \frac{k_i}{k_{n+1}}$$

$$i = 1, \ldots, n$$

For example, in the space of 2-tuples (Fig. 9.5),

$$\mathbf{v}(4,1) = 4\epsilon_1 + \epsilon_2 \qquad \text{(Fig. 9.5}a\text{)}$$

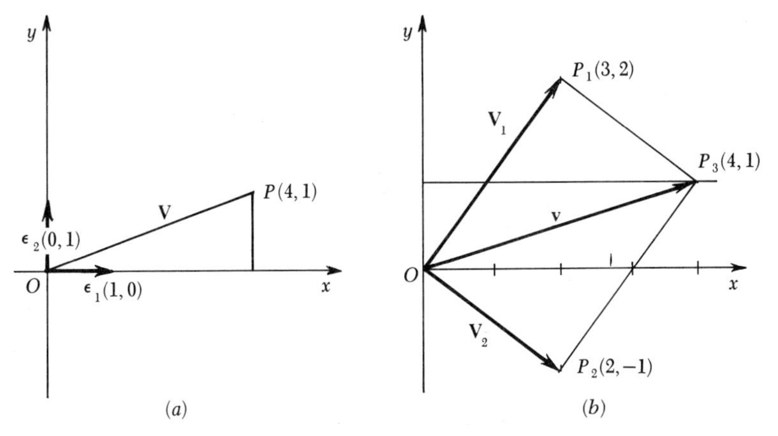

FIG. 9.5

and

$$\mathbf{v}(4,1) = \tfrac{6}{7}\mathbf{v}_1(3,2) + \tfrac{5}{7}\mathbf{v}_2(2,-1) \qquad \text{(Fig. 9.5}b\text{)}$$

where $\mathbf{v}_1(3,2)$ and $\mathbf{v}_2(2,-1)$ are linearly independent vectors.

Each of the sets ϵ_1, ϵ_2 and $\mathbf{v}_1$, $\mathbf{v}_2$ forms a basis for the 2-space in accordance with the definition:

Definition 1 *A set of linearly independent vectors is a basis for a vector space if every vector of the space is a linear combination of the vectors of the set.*

Since in the vector space of n-tuples any $m > n$ vectors are linearly dependent, a basis for the space of n-tuples contains n vectors.

Definition 2 *The number of vectors in a basis for a vector space is the dimension of the space.*

Remark Definition 2 is consistent with the familiar terminology: a line is a one-dimensional space; a plane is a two-dimensional space; ordinary space is three-dimensional, etc. To a point on a line corresponds a single coordinate and hence a vector with one component; to a point in the plane, a pair of numbers and hence a vector with two components; to a point in space, a triple of numbers and hence a vector with three components, etc. However, the definition is not adequate for the dimension of an arbitrary point set such as, for example, the point set R consisting of all points on the x axis whose coordinates are rational numbers. Such matters are treated in texts dealing with the modern theory of dimension—a theory of considerable magnitude.

Exercise 9.3

1 Let $\mathbf{v}_1$ and $\mathbf{v}_2$ be the vectors of Fig. 9.5*b*, and let k be any real number. If

$$\mathbf{w}_1 = \mathbf{v}_1 \qquad \text{and} \qquad \mathbf{w}_2 = k\mathbf{v}_1 + \mathbf{v}_2$$

do $\mathbf{w}_1$ and $\mathbf{w}_2$ form a basis for the space of 2-tuples? Explain.

2 Write three vectors, not the natural vectors, forming a basis for the space of number triples.

3 For what value or values of k will the vectors $(k, 1 - k, k)$, $(2k, 2k - 1, k + 2)$, $(-2k, k, -k)$ form a basis for the vector space of number triples? Explain.

9.5 Geometric meaning of linear dependence

A geometric interpretation of linear dependence will be given next. First, it is noted that the coordinate representation of a point P of the plane by a number triple of the set (ka_1,ka_2,ka_3), with $k \neq 0$, is ambiguous. Sometimes it is necessary to know which number triple of the set is being used in a discussion. When such is the case, a special symbol, explained below, will be used. It is suggested by the fact that a vector in three-dimensional space has three components.

If $\mathbf{v}$ is the vector with components (a_1, a_2, a_3) the symbol

$$P(\mathbf{v})$$

will mean that homogeneous coordinates of P are (a_1, a_2, a_3).

For example, if components of vector $\mathbf{w}$ are $(1,0,-1)$, coordinates of $P(\mathbf{w})$ are $(1,0,-1)$ and of $P(2\mathbf{w})$ are $(2,0,-2)$.

This new symbolism is used in the definitions that follow. They show that operations on points (lines) are analogous to corresponding operations on vectors representing them.

Definition 1 *(Scalar Multiplication)*

$$cP(\mathbf{v}) = P(c\mathbf{v})$$

Multiplication of a point P by a scalar c, therefore, simply multiplies the representative vector of P by c.

Definition 2 *(Addition) If*

$$\mathbf{v} = c_1\mathbf{v}_1 + c_2\mathbf{v}_2 + \cdots + c_m\mathbf{v}_m \neq \theta$$

then

$$c_1P_1(\mathbf{v}_1) + c_2P_2(\mathbf{v}_2) + \cdots + c_mP_m(\mathbf{v}_m) = P(c_1\mathbf{v}_1 + c_2\mathbf{v}_2 + \cdots + c_m\mathbf{v}_m) \quad (1)$$

and

$$\sum_{i=1}^{m} c_1P_i(\mathbf{v}_i) = \theta \quad (2)$$

is taken to mean that

$$\sum_{i=1}^{m} c_i\mathbf{v}_i = \theta$$

For example, $P_i(-5\mathbf{v}_1) + P_2(\mathbf{v}_2) + P_3(-3\mathbf{v}_3) = \theta$ means that

$$-5\mathbf{v}_1 + \mathbf{v}_2 - 3\mathbf{v}_3 = \theta$$

The points P_1, P_2, P_3 are then said to be linearly dependent in accordance with the definition:

Definition 3 *The points $P_i(\mathbf{v}_i)$, with $i = 1, \ldots, m$, are linearly dependent or independent according as the vectors $\mathbf{v}_i$ are linearly dependent or independent.*

From question 6 of Exercise 9.2 it follows that the linear dependence of points is not affected by the particular set of coordinates chosen to represent the points.

Definition 4 *If the points $P_1, P_2, \ldots, P_m$ are linearly dependent, then at least one of the c_i's of (2) is not equal to zero, say c_1, and (2) may be solved for P_1; thus*

$$P_1 = -\frac{1}{c_1}(c_2P_2 + c_3P_3 + \cdots + c_mP_m)$$

The point P_1 is then said to be a *linear combination* of the points $P_2, P_3, \ldots, P_m$.

Each of the definitions 1 to 4 may be dualized by replacing the word "point" by the word "line," and the geometric meaning of linear dependence of points or lines may now be given. This is done in the next set of theorems which follow immediately from the definitions just given and the algebra of Sec. 8.6.

Theorem 9.1 *Two points (lines) are linearly dependent if and only if they coincide.*

Theorem 9.2 *Three points are linearly dependent if and only if they are collinear.*

Theorem 9.3 *Three lines are linearly dependent if and only if they are concurrent.*

Corollary *Theorem 9.2 is illustrated in Fig. 8.6, where the linearly dependent points $A(1,0,-1)$, $B(2,3,1)$, $C(1,2,1)$ are collinear, and Theorem 9.3 in Fig. 8.7, where the linearly dependent lines $a(1,0,-1)$, $b(2,3,1)$, $c(1,2,1)$ are concurrent.*

From Theorems 9.2 and 9.3 and the definition of linear combination follow the dual theorems:

Theorem 9.4 *If A and B are distinct points of the plane, a point C is on the line AB if and only if it is a linear combination of A and B.*

Theorem 9.5 *If a and b are distinct lines of the plane, a line c is on their point of intersection if and only if c is a linear combination of a and b.*

The next theorem follows immediately from the foregoing definitions and Theorem 8.6.

Theorem 9.6 *Any four points (lines) in the projective plane are dependent.*

An interesting fact is observed now. If vectors $\mathbf{v}_1$ and $\mathbf{v}_2$ have respective components $(1,0,-1)$ and $(2,3,1)$,

$$A(\mathbf{v}_1) + 2B(\mathbf{v}_2) = C(5,6,1)$$
$$A(2\mathbf{v}_1) + 2B(\mathbf{v}_2) = D(6,6,0) \neq C$$

and it is seen that the same linear combination $A + 2B$ has yielded two distinct points of the line AB, while different representative vectors have been chosen for the point A.

This ambiguity involved in the coordinate representation of points (lines) can be troublesome at certain times and useful at others.

Particularly useful in this connection is the next theorem:

Theorem 9.7 *If $P_1, P_2, \ldots, P_m$ are linearly independent points while $P_1, P_2, \ldots, P_{m+1}$ are linearly dependent, then coordinates of the points may be chosen so that*

$$P_1 + P_2 + \cdots + P_m = P_{m+1}$$

proof: Since the points $P_1, P_2, \ldots, P_{m+1}$ are linearly dependent, constants $c_1, c_2, \ldots, c_{m+1} \neq 0, 0, \ldots, 0$ exist such that

$$c_1P_1(\mathbf{v}_1) + c_2P_2(\mathbf{v}_2) + \cdots + c_mP_m(\mathbf{v}_m) + c_{m+1}P_{m+1}(\mathbf{v}_{m+1}) = \theta$$

Now $c_{m+1} \neq 0$, for otherwise the points $P_1, \ldots, P_m$ would be dependent contrary to hypothesis. The equation may, therefore, be solved for P_{m+1} giving

$$\begin{aligned} P_{m+1} &= -\frac{1}{c_{m+1}}[c_1P_1(\mathbf{v}_1) + \cdots + c_mP_m(\mathbf{v}_m)] \qquad i = 1, 2, \ldots, m \\ &= k_1P_1(\mathbf{v}_1) + \cdots + k_mP_m(\mathbf{v}_m) = P_1(k_1\mathbf{v}_1) + \cdots + P_m(k_m\mathbf{v}_m) \end{aligned}$$

where $\qquad k_i = \dfrac{c_i}{c_{m+1}}$

or dropping the symbols $k_i\mathbf{v}_i$,

$$P_{m+1} = P_1 + \cdots + P_m$$

and the theorem is proved. A simple illustration follows.

In the equation

$$A(\mathbf{v}_1) + 2B(\mathbf{v}_2) = C(5,6,1)$$

a choice of $2\mathbf{v}_2$ for the representative vector of B gives

$$A + B = C$$

Statement, proof, and illustration of the dual of Theorem 9.7 are left as exercises. (See Exercise 9.4, question 4.)

Exercise 9.4

1 a. If $(2,3,-2)$, $(4,5,2)$, and $(1,2,-4)$ are coordinates of the respective points A, B, C, find constants k, l, m such that

$$kA + lB + mC = \theta$$

b. Show the points in a drawing. Are they collinear? Why?

c. What are the coordinates of the points when

$$A + B + C = \theta$$

d. Dualize (a) and (b).

2 If $[-1,1,2]$, $[-2,2,5]$, $[3,-3,7]$ are coordinates of the three lines a, b, c of the plane, show that representative vectors may be chosen so that

$$a = b + c$$

and constants $k, l \neq 0, 0$ such that

$$ka - lb = c.$$

3 Prove that a necessary and sufficient condition that three distinct lines, one through each vertex of the triangle whose sides are

$$a = 0 \qquad b = 0 \qquad c = 0$$

be concurrent is that their equations may be put into the forms

$$kb = lc \qquad lc = ma \qquad ma = kb \qquad k, l, m \neq 0, 0, 0$$

4 State, prove, and illustrate the dual of Theorem 9.7.

9.6 Geometrical applications of linear dependence

The time has come now to fulfill an earlier promise that analytic proofs would replace some earlier ones. Synthetic proofs often require some special skill, insight, or ability and sometimes require just luck. The new proofs will show how much the reasoning process may be mechanized.

First to be proved is the first half of Desargues' theorem:

Theorem 9.8 *If two triangles ABC and $A'B'C'$ are perspective from a point O, they are perspective from a line.*

proof: Let O be the point of meeting of the lines AA', BB', CC' joining corresponding vertices of the two triangles. Then, since O is on each of the lines AA', BB', CC', scalars a, b, c exist such that

$$O = aA + a'A' = bB + b'B' = cC + c'C'$$

and hence

$$aA - bB = a'A' - bB' \tag{1}$$
$$bB - cC = b'B' - cC' \tag{2}$$
$$cC - aA = a'C' - a'A' \tag{3}$$

Now, the left hand side of (1) represents a point C'' on line AB, and the right side a point on line $A'B'$. Since the two points are equal, C'' is also on line $A'B'$. In other words, C'' is the point of intersection of corresponding sides AB and $A'B'$. Similarly, A'' is the point of intersection of sides BC and $B'C'$, and B'' of sides AC and $A'C'$. But then

$$aA - bB + bB - cC + cC - aA = \theta$$

or equivalently,

$$C'' + B'' + A'' = \theta$$

and by Theorem 9.2, the three points A'', B'', C'' are collinear.

A reversal of these arguments gives a proof of the converse (dual) theorem:

Theorem 9.9 *If two triangles are perspective from a line, they are perspective from a point.*

The absence of a reference figure in the proof was deliberate. The reader should decide for himself the advantages and disadvantages (if any) of a proof based on purely algebraic processes. He should also compare this new proof with the earlier one which required separate treatments for the plane and 3-space—and which was considerably more complicated. The analytic proof given here requires no special concern with figures and extends easily to spaces of higher dimensions.

The theorem stated and proved next will be recognized as an earlier axiom (Axiom 9, Sec. 5.7), and this time a figure will be used to illustrate the algebraic processes.

Theorem 9.10 *The diagonal points of a complete quadrangle are not collinear.*

proof: Let P_i, $i = 1, \ldots, 4$, be the four vertices of a quadrangle, and let D_j, $j = 1, 2, 3$, be its diagonal points (Fig. 9.6).

Since no three vertices are collinear, it follows from Theorem 9.7 that coordinates of the vertices may be chosen so that

$$P_1 + P_2 + P_3 + P_4 = \theta$$

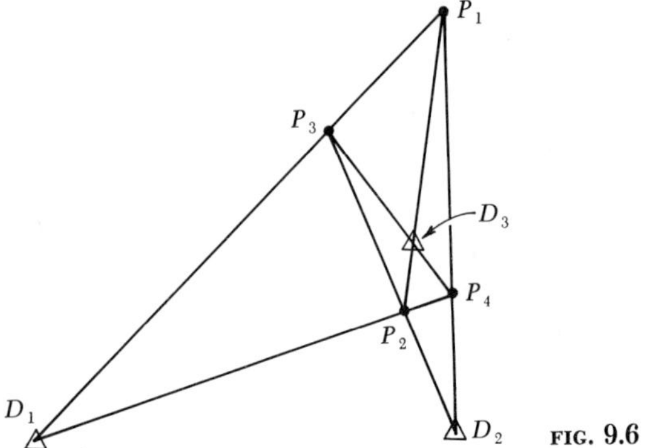

FIG. 9.6

and hence

$$P_1 + P_2 = -P_3 - P_4 \quad (1)$$
$$P_2 + P_3 = -P_1 - P_4 \quad (2)$$
$$P_3 + P_2 = -P_1 - P_4 \quad (3)$$

Equation (1) shows that a point on line P_1P_2 is also a point on line P_3P_4 and is therefore a diagonal point, say D_3. Similarly, from (2) lines P_2P_3 and P_4P_1 meet in a second diagonal point D_2, and from (3) lines P_3P_1 and P_2P_4 meet in the third diagonal point D_1. The diagonal points satisfy the relations

$$D_1 = P_3 + P_1 \quad (4)$$
$$D_2 = P_2 + P_3 \quad (5)$$
$$D_3 = P_1 + P_2 \quad (6)$$

The theorem is then proved by showing that the equation

$$k_1D_1 + k_2D_2 + k_3D_3 = \theta \quad (7)$$

is satisfied only for $k_1 = k_2 = k_3 = 0$.

Substitution in this equation of the values for D_1, D_2, D_3 in (4), (5), and (6) gives

$$(k_2 + k_3)P_1 + (k_1 + k_3)P_2 + (k_1 + k_3)P_3 = \theta$$

and because P_1, P_2, P_3 are linearly independent, k_1, k_2, k_3 must satisfy the system of equations

$$k_2 + k_3 = 0$$
$$k_1 + k_3 = 0$$
$$k_1 + k_2 = 0$$

with the unique solution $k_1 = k_2 = k_3 = 0$. The points D_1, D_2, D_3 are, therefore, linearly independent, or in other words, not collinear. Thus, the theorem is proved.

Exercise 9.5

1 State and prove by analytic methods the dual of Theorem 9.10.

2 Prove that the three points in which nonconcurrent sides of a complete quadrangle, one through each diagonal point, meet opposite sides of the diagonal triangle are collinear.

3 By analytic methods, prove Pappus' theorem: "If distinct points A, B, and C lie on a line and if distinct points A', B', and C' lie on a second distinct, coplanar line, the points of intersection of AB' and $A'B$, of AC' and $A'C$, and of BC' and $B'C$ are collinear." *Hint:* See ([50] Sec. 8.2).

9.7 Cross ratio in coordinate geometry

When cross ratio was introduced earlier,[1] metric concepts were used and coordinates had not yet been introduced. A coordinate formula for cross ratio in the projective plane is to be obtained here by modifying a corresponding coordinate formula for cross ratio in the Euclidean plane.

If the coordinate y_i of a point P_i of the Euclidean line is its distance from a fixed point of the line, the early formula for cross ratio

$$\lambda = (P_1P_2,P_3P_4) = \frac{P_1P_3/P_3P_2}{P_1P_4/P_4P_2} \tag{1}$$

may be changed, by the substitution $P_iP_j = y_j - y_i$, to the coordinate formula

$$\lambda = \frac{(y_3 - y_1)(y_2 - y_4)}{(y_2 - y_3)(y_4 - y_1)} \tag{2}$$

or by the substitution $y_i = x_i/t_i$, to the formula

$$\lambda = (P_1P_2,P_3P_4) = \frac{\begin{vmatrix} x_3 & t_3 \\ x_1 & t_1 \end{vmatrix}\begin{vmatrix} x_2 & t_2 \\ x_4 & t_4 \end{vmatrix}}{\begin{vmatrix} x_2 & t_2 \\ x_3 & t_3 \end{vmatrix}\begin{vmatrix} x_4 & t_4 \\ x_1 & t_1 \end{vmatrix}} \tag{3}$$

By definition, the cross ratio $\lambda = (P_1P_2,P_3P_4)$ of the four points $P_i(x_i,t_i)$, $i = 1, \ldots, 4$ of the projective line is the quantity λ given by (3). Dually, if $[x_i,t_i]$ are homogeneous coordinates of a line L_i of the projective plane, the

[1] See Sec. 4.10.

0 $P_1(1)$ $P_3(2)$ $P_2(3)$

FIG. 9.7

cross ratio $\lambda = (L_1L_2,L_3L_4)$ of the four lines L_i, $i = 1, \ldots, 4$, on a point is given by (3).

Formulas (2) and (3) are applied and contrasted. Let P_1, P_2, P_3 be three collinear points whose nonhomogeneous coordinates 1, 2, 3 are respective distances of the points from a fixed point O of their line (Fig. 9.7). It is desired to find a fourth point $P_4(y_4)$ such that

$$\lambda = (P_1P_2,P_3P_4) = -1$$

(P_4 is then the harmonic conjugate of P_3 with respect to P_1 and P_2.)

Now, from (1),

$$\lambda = \frac{(2-1)(3-y_4)}{(3-2)(y_4-1)} = -1$$

and hence

$$3 - y_4 = -y_4 + 1 \qquad \text{or} \qquad 3 = 1$$

This inconsistent result means, in geometric terminology, that on the Euclidean line there is *no harmonic conjugate to the midpoint of a line segment* P_1P_2 *with respect to the points* P_1, P_2.

Contrast this result with that obtained by using formula (3) with (1,1), (2,1), (3,1) as homogeneous coordinates for the same three points P_1, P_2, P_3. Then

$$\lambda = \frac{\begin{vmatrix} 2 & 1 \\ 1 & 1 \end{vmatrix} \begin{vmatrix} 3 & 1 \\ x_4 & t_4 \end{vmatrix}}{\begin{vmatrix} 3 & 1 \\ 2 & 1 \end{vmatrix} \begin{vmatrix} x_4 & t_4 \\ 1 & 1 \end{vmatrix}} = -1$$

and hence $t_4 = 0$. This time, the desired point P_4 has been found. It is the ideal point of the *projective line.* One cannot fail to be impressed by the projective theory that has generalized elementary Euclidean theory.

Some other facts are noted. A simple calculation shows that substitution in (3) of (k_ix_i,k_it_i), $k_i \neq 1$, for (x_i,t_i) does not change the value of the cross ratio λ. This fact is stated formally in the theorem:

Theorem 9.11 *The cross ratio λ of four points (lines) is independent of the particular pair of coordinates of the set (kx_i,kt_i) used to represent an element.*

The restriction is removed now that the four elements in question are *distinct.* Suppose that one of them, say P_4, coincides with one of the remaining three distinct elements. What new values of the cross ratio are then possible?

A reference to (3) shows that when P_4 *coincides with* P_3, numerator and denominator are alike and hence

$$\lambda = (P_1P_2,P_3P_3) = 1$$

If P_4 *coincides with* P_2,

$$\begin{vmatrix} x_2 & t_2 \\ x_4 & t_4 \end{vmatrix} = 0$$

and

$$\lambda = (P_1P_2,P_3P_2) = 0$$

Finally, if P_4 *coincides with* P_1,

$$\begin{vmatrix} x_4 & t_4 \\ x_1 & t_1 \end{vmatrix} = 0$$

and

$$\lambda = (P_1P_2,P_3P_1) = \infty$$

Thus, there has been admitted to the set of possible values of the cross ratio of four elements, the additional values 1, 0, and ∞.

Exercise 9.6

1 Given the collinear points $P_1(1)$, $P_2(4)$, $P_3(7)$, $P_4(10)$ and the collinear points $P'_1(0)$, $P'_2(1)$, $P'_3(\frac{4}{3})$, $P'_4(\frac{3}{2})$, show that the cross ratios (P_1P_2,P_3P_4) and $(P'_1P'_2P'_3P'_4)$ are equal.

2 If E_1, E_2, E_3, E_4 are distinct elements of a pencil of points (pencil of lines) whose homogeneous coordinates are, respectively,

$$(a_i) \qquad (b_i) \qquad (a_i + \lambda_1 b_i) \qquad (a_i + \lambda_2 b_i) \qquad i = 1, \ldots, 4$$

show that $(E_1,E_2,E_3,E_4) = \lambda_1/\lambda_2$.

3 If A, B, C are any three points on the line through two given points O and U, show that the cross ratios (OU,AB), (OU,BC), and (OU,AC) satisfy the relation

$$(OU,AB) \cdot (OU,BC) = (OU,AC)$$

Concluding remarks

This chapter has been devoted to showing the usefulness of vectors and vector theory in algebraic geometry. It also emphasizes the close relationship between geometry and basic concepts of linear algebra. In fact, these two branches of mathematics, geometry on the one hand and linear algebra on the other, are often said to be but two sides of the same coin.

Among other things, linear dependence has been used to introduce a *basis* for a vector space, and the latter, in turn, to define the dimension of a space. Linear dependence has also been used to give neat, concise, condensed proofs and generalizations of earlier geometric theory. Representing coordinates of a point or line by components of a vector has many advantages: It saves time and ink, it clarifies, it generalizes. What is more, proofs given here can be extended easily to spaces of a higher dimension than three ([113] Chap. 5).

The new formulas for cross ratio in terms of coordinates are for use in the next chapter, where the dominant role of this quantity is reemphasized.

A reference to some of the older literature ([53] Chap. 7) will show the advantages of a vector approach to analytic geometry and the almost electrifying effect that vector terminology and symbolism have on classical theory. Vectors and matrices are as valuable to the modern geometer as modern tools are to technicians of the Jet Age.

Suggestions for further reading

Fishbach, W. T.: "Projective and Euclidean Geometry," chap. 7.
Graustein, W. E.: "Introduction to Higher Geometry," chap. 3.
Hausner, Melvin: "A Vector Space Approach to Geometry," chaps. 2 and 3.
Paige, Lowell J., and J. Dean Swift: "Elements of Linear Algebra," chaps. 2 and 3.
Pedoe, Daniel: "A Geometric Introduction to Linear Algebra."
Robinson, G. de B.: "Vector Geometry," chap. 5.
Rosenbaum, Robert A.: "Introduction to Projective Geometry and Modern Algebra," chap. 4.

10 generalized coordinate systems and linear transformations

There is a process in elementary analytic geometry called *change, or transformation, of coordinates* by virtue of which a point P of the plane acquires a set of coordinates for each position of a reference system. For instance, in Fig. 10.1 point P has rectangular coordinates x, y relative to the axes intersecting at O and rectangular coordinates x', y' relative to those intersecting at point $O'(h,k)$. The two sets of coordinates are related by the system of equations

$$\begin{aligned} x &= x' + h \\ y &= y' + k \end{aligned} \tag{1}$$

One original purpose in changing reference axes in the Euclidean plane was to reduce the algebraic complexities of a problem. For example, a rotation of axes through 45° followed by a translation to the new origin $O'(-\frac{1}{2}\sqrt{2},\frac{7}{2}\sqrt{2})$ of Fig. 10.2 reduces the equation

$$5x^2 + 6xy + 5y^2 + 22x - 6y + 21 = 0 \tag{2}$$

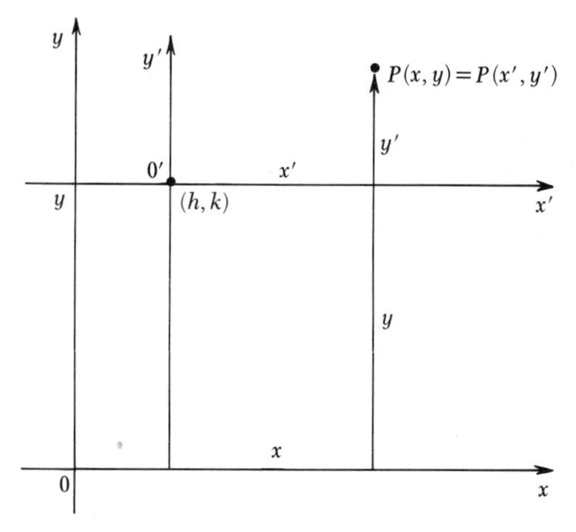

FIG. 10.1

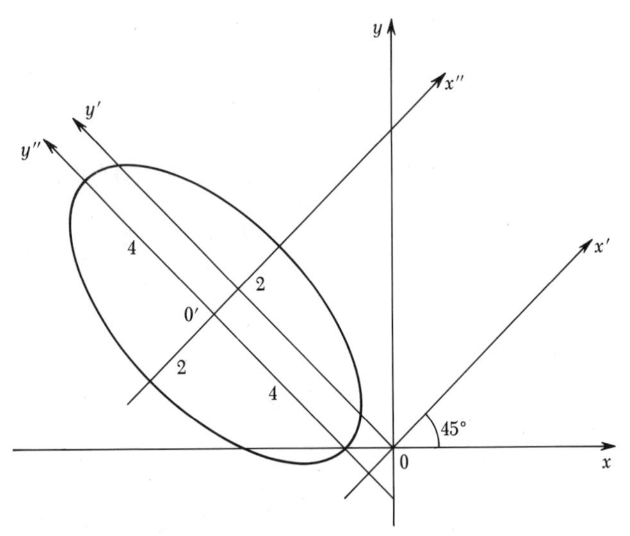

FIG. 10.2

to the simple form

$$4x''^2 + y''^2 = 16 \tag{3}$$

showing clearly that the curve is an ellipse with minor axis of length 4 on the x'' axis, and major axis of length 8 on the y'' axis.

However, system (1) may be given quite a different interpretation. The equations may be viewed as machinery that sends or moves a point $P(x,y)$ into a new point $P'(x',y')$, *the axes remaining fixed.* In more formal language they are mappings of points into points.

The use of the word "move" in this connection calls to mind some of the earlier controversies concerning *motion* and *congruency.*[1] According to Euclid, two triangles were congruent when one could be *moved* so as to coincide with the other. This is a queer mixture of the physical with the abstract. A triangular piece of cardboard can be moved and the motion pictured, as in Fig. 10.3. But actually, *points don't move.* The figures should be interpreted to mean simply that under the particular translation

$$x' = x + 4 \qquad y' = y - 2$$

points

$$A(2,0) \qquad B(4,1) \qquad C(2,1)$$

[1] See Sec. 2.9.

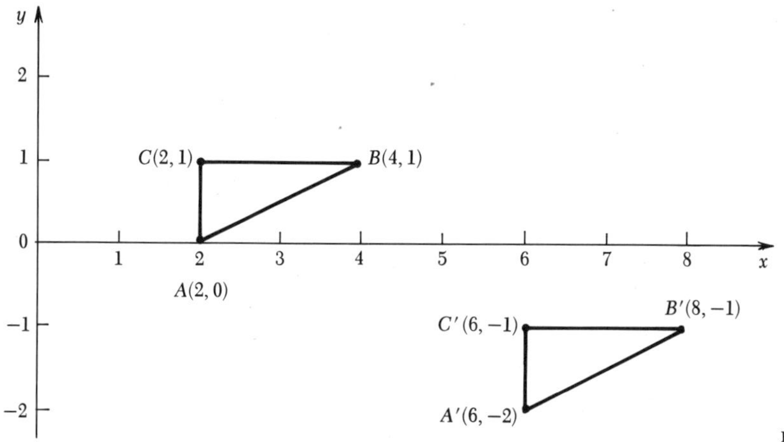

FIG. 10.3

are mapped into the points

$$A'(6,-2) \qquad B'(8,-1) \qquad C'(6,-1)$$

The fact that triangle ABC and its transform, triangle $A'B'C'$, are congruent is not because one has been moved to another position but because the particular transformation had congruency as an invariant.

Both ways of looking at system (1) are significant. It is repeated and emphasized that when the system is viewed as a mapping, *the point* $P(x,y)$ *is moved,* so to speak, to another point P' (Fig. 10.4), *the reference axes remaining fixed.* When the system is viewed as a change of coordinate formulas, *the point* $P(x,y)$ *remains fixed, and the reference axes are moved* (Fig. 10.1).

Setting up different coordinate systems in the projective plane is not as simple as it was in the Euclidean plane. Linear dependence replaces

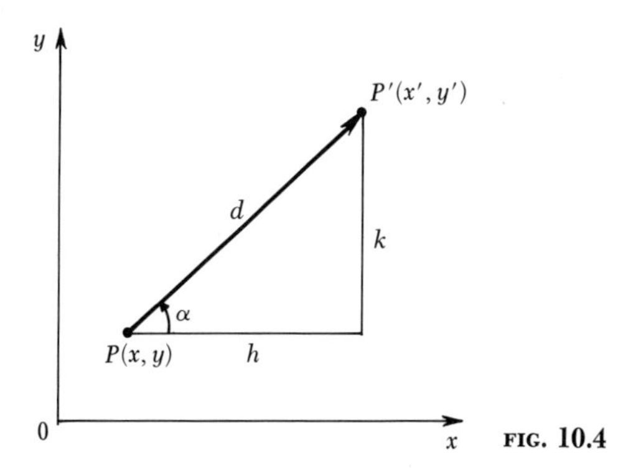

FIG. 10.4

the metric concepts used there, and the results are of far greater importance than the reduction of algebraic complexities of a problem.

The change of coordinate process in the projective plane will be carried out here, and the formulas derived in the process will be viewed as transformations or mappings. Interest will shift then to properties of figures invariant under these transformations. Discovery of the invariance of cross ratio was one of the highlights of the search for invariants.

One of the primary aims of this chapter is an introduction to the Erlanger program, proposed by Klein in 1872 when he delivered his inaugural address to the Philosophical Faculty and the Senate of the University of Erlanger. The program classifies geometries by means of their invariants under groups of transformations and is in no way at variance with the axiomatic classification theories presented in Chap. 2.

The Erlanger program is simply another method which neatly catalogues, classifies, and unifies existing theory. Despite some of its limitations, Klein's very ingenious program has had a profound effect on all subsequent thinking.

10.1 Generalized coordinate systems in the plane. Change of coordinate formulas

It is proposed now to generalize the natural homogeneous coordinate system of Chap. 8 and then to set up change of coordinate formulas for use in later theory. As will be seen presently, linear dependence figures prominently in the process and takes the place of the metric concepts used in the corresponding problem in the Euclidean plane.

In the projective plane, the familiar x, y reference axes of a rectangular coordinate system are replaced by a reference triangle formed by three linearly independent points.

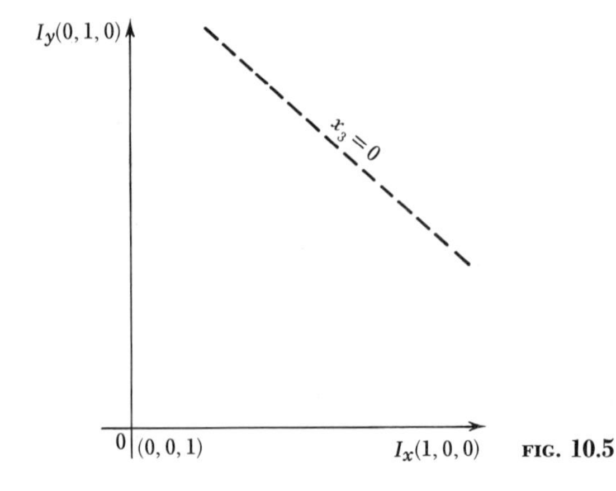

FIG. 10.5

A reference triangle most closely linked with the natural homogeneous coordinates is shown in Fig. 10.5 with two of its sides coinciding with the horizontal and vertical lines through point O and its third side $x_3 = 0$, the (dotted) line joining the ideal point I_x on the horizontal line to the ideal point I_y on the vertical line.

If ϵ_1, ϵ_2, ϵ_3 are the three linearly independent natural 3-dimensional vectors

$$\epsilon_1(1,0,0) \qquad \epsilon_2(0,1,0) \qquad \epsilon_3(0,0,1)$$

the vertices of this reference triangle are

$$I_x(\epsilon_1) \qquad I_y(\epsilon_2) \qquad O(\epsilon_3)$$

Suppose now that the representative vector $\mathbf{v}$ of another point P of the plane has components x_1, x_2, x_3. Then, since the four points P, I_x, I_y, O are *linearly dependent*, $P(\mathbf{v})$ may be expressed

$$P(\mathbf{v}) = x_1 I_x(\epsilon_1) + x_2 I_y(\epsilon_2) + x_3 O(\epsilon_3) \tag{1}$$

and in this form, homogeneous coordinates of P are seen *to be scalar multiples of the vertices I_x, I_y, and O.* The numbers x_1, x_2, x_3 are therefore called *homogeneous coordinates of point P with respect to the reference triangle I_xI_yO.*

It seems quite natural to generalize this situation by replacing the particular points I_x, I_y, O by any three linearly independent points

$$P_1(\mathbf{v}_1) \qquad P_2(\mathbf{v}_2) \qquad P_3(\mathbf{v}_3)$$

expressing $P(\mathbf{v})$ as

$$P(\mathbf{v}) = \lambda P_1(\mathbf{v}_1) + \mu P_2(\mathbf{v}_2) + \gamma P_3(\mathbf{v}_3) \tag{2}$$

and calling

$$\lambda \qquad \mu \qquad \gamma$$

coordinates of point P *relative to the reference triangle $P_1P_2P_3$.*

Unfortunately, this is not a satisfactory program, for a change in the representative vector of a point P from $\mathbf{v}$ to $\mathbf{v}/2$ with no change in representatives of the other points gives

$$P(\mathbf{v}) = 2\lambda P_1\left(\frac{\mathbf{v}_1}{2}\right) + \mu P_2(\mathbf{v}_2) + \gamma P_3(\mathbf{v}_3)$$

and this time, the number triple associated with point P is

$$(2\lambda,\mu,\gamma)$$

A number triple that can be varied arbitrarily cannot serve as a set of coordinates for a point. However, there is a way out of the difficulty. Representative vectors $\mathbf{v}_i'$ of points $\mathbf{P}_i(\mathbf{v}_i)$ are determined by the agreement that a point $U(\mathbf{a})$ is to be a unit point (1,1,1) or (k,k,k) in a coordinate system having $P_1P_2P_3$ as a reference triangle.

Now, if

$$\begin{aligned} \mathbf{a} &= c_1\mathbf{v}_1 + c_2\mathbf{v}_2 + c_3\mathbf{v}_3 \\ U(\mathbf{a}) &= \sum_{i=1}^{3} P_i(c_i\mathbf{v}_i) \end{aligned} \tag{3}$$

and if

$$\begin{aligned} \mathbf{v}_i' &= c_i\mathbf{v}_i \\ U(\mathbf{a}) &= 1 \cdot P_1(\mathbf{v}_1') + 1 \cdot P_2(\mathbf{v}_2') + 1 \cdot P_3(\mathbf{v}_3') \end{aligned} \tag{4}$$

The scalar multiples 1, 1, 1 of points P_i in (4) are homogeneous coordinates of point U, and hence U is a unit point. Moreover, if $\mathbf{U}(\mathbf{a})$ is to retain its character of being a unit point, any change in a representative vector of a point, say P_1, from $\mathbf{v}_1'$ to $k\mathbf{v}_1'$ must be accompanied by corresponding changes in points P_2 and P_3, for then

$$U(k\mathbf{a}) = k \sum_{i=1}^{3} P_i(\mathbf{v}_i')$$

and U is still the unit point (k,k,k).

Suppose now that components of the respective vectors $\mathbf{v}_1'$, $\mathbf{v}_2'$, $\mathbf{v}_3'$ are given by the number triples

$$(a_{11},a_{12},a_{13}) \qquad (a_{21},a_{22},a_{23}) \qquad (a_{31},a_{32},a_{33})$$

and that the equation expressing $P(k\mathbf{v})$ in terms of P_1, P_2, P_3 is

$$P(k\mathbf{v}) = x_1'P_1(\mathbf{v}_1') + x_2'P_2(\mathbf{v}_2') + x_3'P(\mathbf{v}_3') \tag{5}$$

Then (5) yields the system of equations

$$\begin{aligned} kx_1 &= a_{11}x_1' + a_{21}x_2' + a_{31}x_3' \\ kx_2 &= a_{12}x_1' + a_{22}x_2' + a_{32}x_3' \\ kx_3 &= a_{13}x_1' + a_{23}x_2' + a_{33}x_3' \end{aligned} \tag{6}$$

which reduces, in matrix form, to

$$kX = AX'$$

where X, A, X' are the matrices

$$X = \begin{pmatrix} x_1 \\ x_2 \\ x_3 \end{pmatrix} \qquad A = \begin{pmatrix} a_{11} & a_{21} & a_{31} \\ a_{12} & a_{22} & a_{32} \\ a_{13} & a_{23} & a_{33} \end{pmatrix} \qquad X' = \begin{pmatrix} x_1' \\ x_2' \\ x_3' \end{pmatrix}$$

Since the points P_1, P_2, P_3 are linearly independent, matrix A has an inverse A^{-1}, and system (6) may be solved for x_1', x_2', x_3' in terms of kx_1, kx_2, kx_3. In matrix form the solution is

$$X' = kA^{-1}X \tag{7}$$

By definition, the numbers x_1', x_2', x_3' satisfying (7) are homogeneous coordinates of point $P(\mathbf{v})$ in a coordinate system having $P_1P_2P_3$ as the reference triangle and $U(\mathbf{a})$ as the unit point.

The results just established are stated formally in the theorem:

Theorem 10.1 *A coordinate system for points of the projective plane is uniquely determined by four points, no three of which are collinear. Three of them are vertices of a reference triangle, and the fourth is the unit point.*

Remark 1 Note is made in passing that the natural coordinates x_1, x_2, x_3 of point $P(\mathbf{v})$ are simply coordinates relative to the system having points

$$I_x(\epsilon_1) \qquad I_y(\epsilon_2) \qquad O(\epsilon_3)$$

as vertices of the reference triangle and $U(1,1,1)$ as the unit point.

Remark 2 System (7) associates with each point P of the plane a new set of coordinates x_1', x_2', x_3', and a simple calculation (see Exercise 10.1, question 1) shows that when $P = P_1(a_{11},a_{12},a_{13})$, the new number triple representing P_1 is (1,0,0). Similarly, new coordinates of P_2 are (0,1,0), and of P_3 are (0,0,1).

Fig. 10.6 shows both the old, and beneath them, the new coordinates of points P_1, P_2, P_3 and brings out graphically the fact that a point $P(x_1,x_2,x_3)$, with $x_3 = 0$, is no longer the special point called in Chap. 8 the infinitely distant point of the line.

An example of how four points are used to set up a coordinate system follows.

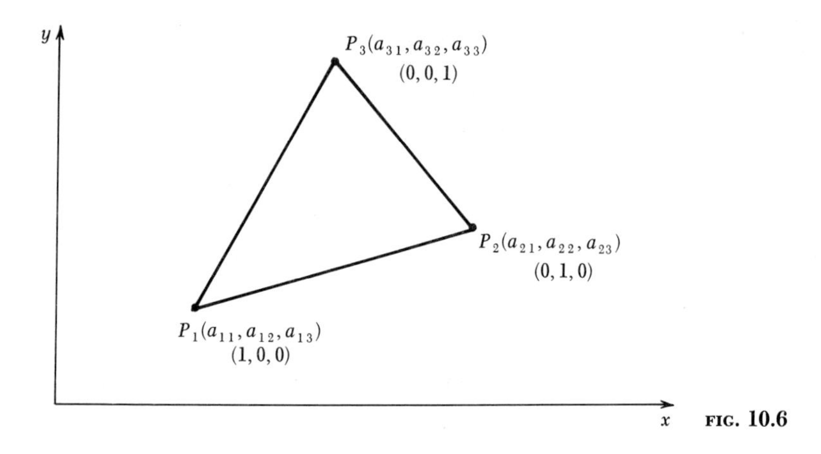

FIG. 10.6

Suppose that the four points P_i, with $i = 1, \ldots, 4$, are, respectively,

$$P_1(1,1,0) \qquad P_2(1,1,2) \qquad P_3(0,1,4) \qquad P_4(5,9,22)$$

and $P(2,-3,4)$ is a fifth point. It is desired to find coordinates of point P *relative to the system* in which P_1, P_2, P_3 are vertices *of the reference triangle* and P_4 is the unit point.

Since

$$\mathbf{u}(5,9,22) = 2\mathbf{v}_1(1,1,0) + 3\mathbf{v}_2(1,1,2) + 4\mathbf{v}_3(0,1,4)$$

suitable coordinates for P_1, P_2, P_3 are

$$(2,2,0) \qquad (3,3,6) \qquad (0,4,16)$$

and system (6) then reduces to the system

$$\begin{aligned} kx_1 &= 2x_1' + 3x_2' + 0x_3' \\ kx_2 &= 2x_1' + 3x_2' + 4x_3' \\ kx_3 &= 0x_1' + 6x_2' + 16x_3' \end{aligned} \tag{8}$$

which, for $(x_1,x_2,x_3) = (2,-3,4)$, has the solution

$$x_1' = 10k \qquad x_2' = -8k \qquad x_3' = \tfrac{5}{2}k \qquad k \neq 0$$

A representative set of coordinates for $P(2,-3,4)$ in the new coordinate system is, therefore,

$$(20,-16,5)$$

Dualization of this example and the theory of this section is left as an exercise (See Exercise 10.1, question 2.)

Exercise 10.1

1 a. Write in determinant form the solution of system (6) of Sec. 10.1 for x_1', x_2', and x_3' in terms of x_1, x_2, and x_3.
b. Find a representative set of values of x_1', x_2', x_3' when $(x_1,x_2,x_3) = (a_{11},a_{12},a_{13})$, then (a_{21},a_{22},a_{23}), and finally (a_{31},a_{32},a_{33}).

2 Show in a drawing the lines $p_1[1,1,0]$, $p_2[1,1,2]$, $p_3[0,1,4]$, and $p_4[5,9,22]$, and find coordinates of line $[2,-3,4]$ in a coordinate system having p_1, p_2, p_3 as sides of a reference triangle and p_4 as unit line.

3 Given the points $P_1(1,1,0)$, $P_2(1,0,1)$, $P_3(0,1,1)$, $U(3,-2,-1)$, where the number triple associated with a point represents its natural coordinates:
a. Which one of the points is an ideal point, i.e., an infinitely distant point? Why? Show the other points in an x, y rectangular coordinate system.
b. If P_1, P_2, P_3 are vertices of a reference triangle and U the unit point of a 2nd coordinate system II, write the matrix A of the system which expresses new coordinates x_1', x_2', x_3' of a point P in terms of its original coordinates x_1, x_2, x_3.
c. Calculate the inverse matrix A^{-1} of A.
d. Find coordinates of point $P(2,-3,1)$ with respect to system II. [Answer: $(-6, 9,4)$.]

4 a. If a change of coordinates from x_1, x_2, x_3 to x_1, x_2', x_3 is given by the matrix equation

$$KX = AX'$$

where A is the nonsingular matrix of system (6) (Sec. 10.1), and if a change from x_1', x_2', x_3' to x_1'', x_2'', x_3'' is given by the matrix equation

$$K'X' = BX''$$

where B is the matrix

$$B = \begin{pmatrix} b_{11} & b_{12} & b_{13} \\ b_{21} & b_{22} & b_{23} \\ b_{31} & b_{32} & b_{33} \end{pmatrix}$$

show that a direct change from x_1, x_2, x_3 to x_1'', x_2'', x_3'' is given by the matrix equation

$$K''X = ABX''$$

b. Is the matrix AB singular or nonsingular? Wny?

10.2 Generalized coordinate systems on a line imbedded in a given plane

Within the two-dimensional space (a plane) there are the one-dimensional spaces (lines). Only slight modifications of the preceding theory are needed to set up general coordinate systems for points on a line of the plane or its dual, coplanar lines on a point.

First it is noted that in an $n \geq 1$ projective space $n + 1$ numbers represent (determine) a point (or, in the dual theory, a line). Thus, corresponding to each element of $n = 1$ space is a pair of numbers. The symbol $P(\mathbf{v})$ is used to denote the number pair (x_1x_2) where components of **v** are (x_1,x_2).

If (kx_1,kx_2) denote a point P of the projective line through the fixed points P_1 and P_2, then, as in earlier theory there exist constants (x_1',x_2') satisfying the relation

$$P = x_1'P_1 + x_2'P_2 \tag{1}$$

As in the foregoing section, the number pair x_1', x_2' thus associated with point P is uniquely determined only when representative vectors of P_1 and P_2 have been so chosen that a third point of the line is a unit point. This means that, for points of a line, Theorem 10.1 is replaced by the theorem:

Theorem 10.2 *A coordinate system for points of a line (a pencil of points) is completely determined by three points. Two of them are reference points and the third is the unit point of the system.*

If coordinates of P_1 and P_2 are respectively (a_{11},a_{12}) and (a_{21},a_{22}) and of P are (kx_1,kx_2) there follows from (1) the system of equations

$$\begin{aligned} kx_1 &= a_{11}x_1' + a_{12}x_2' \\ kx_2 &= a_{21}x_1' + a_{22}x_2' \end{aligned} \tag{2}$$

relating original coordinates $\{kx_1,kx_2\}$ of point P to a new set x_1', x_2' relative to the system having P_1 and P_2 as reference points and a third point of the line as a unit point.

Dualization of these results is left as an exercise and attention is directed now to a example illustrating the theory for a particular line of the plane. The example will emphasize the fact that a triple of numbers determine a point of the plane whereas the same point is determined by a pair of numbers when P is specified as a point of a given line of the plane.

Let $\mathbf{a}_1(1,0,-1)$, $\mathbf{a}_2(2,3,1)$, and $\mathbf{c}(-1,1,2)$ be representative vectors of points A_1, A_2, and C of the plane (Fig. 10.7).

A coordinate system is set up on line A_1A_2 having point C as the unit point U.

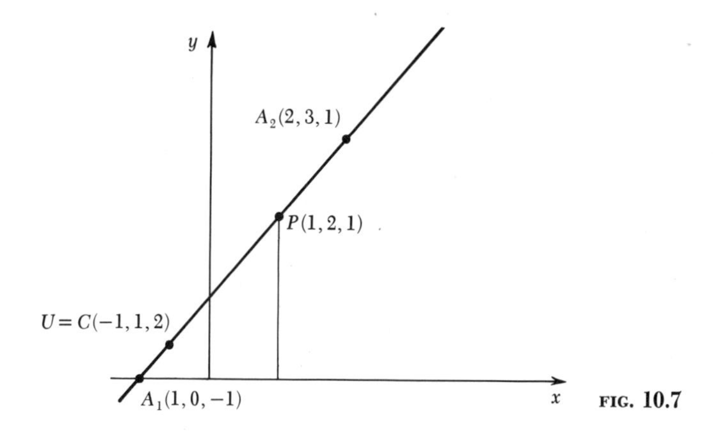

FIG. 10.7

Since

$$\mathbf{u} = \frac{-5}{3}\mathbf{a}_1 + \frac{1}{3}\mathbf{a}_2$$

suitable coordinates for A_1 and A_2 are

$$a' = -\frac{5}{3}a_1 \qquad a_2' = \frac{1}{3}a_2$$

Then, because the points $P_1A_1A_2$ are collinear, and hence linearly dependent, there exist constants y_1 and y_2 satisfying the condition:

$$P(k,2k,k) = y_1A_1(\mathbf{a}_1') + y_2A_2(\mathbf{a}_2')$$

The numbers y_1 and y_2 therefore satisfy the system of equations

$$\frac{-5}{3}y_1 + \frac{2}{3}y_2 = k$$

$$y_2 = 2k$$

$$\frac{5}{3}y_1 + \frac{1}{3}y_2 = k$$

with solution:

$$y_1 = \frac{k}{5} \qquad y_2 = 2k$$

Any pair of numbers of the set $\{k/5,2k\}$ are therefore homogeneous coordinates of the point $P_1(k,2k,k)$ of line A_1A_2.

The single number

$$\lambda = \frac{x_2}{x_1} = \frac{2k}{k/5} = 10$$

is then the nonhomogeneous coordinate of P_1.

Dually, if $a[1,0,-1]$ and $b[2,3,1]$ (Fig. 8.7) are reference lines of the pencil at $P(1,-1,1)$ and if the unit line is taken to be $c[-1,1,2]$, homogeneous coordinates of line $[k,2k,k]$ as a line of the pencil at P are

$$\left(\frac{k}{5},2k\right)$$

and $2k/(k/5) = 10$ is its homogeneous coordinate.

Exercise 10.2

1 Derive system (1) of Sec. 10.2.

2 Find a number pair representing homogeneous coordinates of point $D(1,2,1)$ on the line $A(1,0,-1)$ $B(2,3,1)$ (Fig. 8.6) containing the point $C(-1,1,2)$, if A and C are *reference points* and B is the *unit point.*

3 State and prove the dual of Theorem 10.2.

4 Let lines $a[1,0,-1]$, $b[2,3,1]$, $c[-1,1,2]$ be the respective reference lines and unit line of a coordinate system for the pencil of lines through the point $P(1,-1,1)$ (Fig. 8.7). Find (a) a pair of numbers representing homogeneous coordinates and (b) a single number representing the nonhomogeneous coordinate of the line $[1,2,1]$ of the pencil.

10.3 Sets, groups, and subgroups of linear transformations

The two systems of equations

$$\begin{aligned} x_1' &= a_{11}x_1 + a_{12}x_2 \\ x_2' &= a_{21}x_1 + a_{22}x_2 \end{aligned} \tag{1}$$

$$\begin{aligned} x_1' &= a_{11}x_1 + a_{12}x_2 + a_{13}x_3 \\ x_2' &= a_{21}x_1 + a_{22}x_2 + a_{23}x_3 \\ x_3' &= a_{31}x_1 + a_{32}x_2 + a_{33}x_3 \end{aligned} \tag{2}$$

representing a change of coordinates in one- and two-dimensional space are viewed now as mappings, or transformations.

Because all variables appearing in these systems are raised to only the first power, the systems are called *linear transformations.* Because both systems map *points on a line into points on a line* they are also known as *collineations.*

If no restriction is placed on the matrices of (1) and (2), each system represents a set of transformations; but if the matrices are restricted to be nonsingular, each set has the group property, where the operation of the group is one transformation followed by another. The last statement follows immediately from the corresponding group property of nonsingular $n \times n$ matrices (Exercise 8.6, question 8).

In nonhomogeneous coordinates, (1) takes the form

$$x' = \frac{a_{11}x + a_{12}}{a_{21}x + a} \tag{3}$$

where $x_1'/x_2' = x'$ and $x_1/x_2 = x$. Similarly (2) takes the form:

$$x' = \frac{a_{11}x + a_{12}y + a_{13}}{a_{31}x + a_{32}y + a_{33}}\ y' = \frac{a_{21}x + a_{22}y + a_{23}}{a_{31}x + a_{32}y + a_{33}} \tag{4}$$

which for special values of the coefficients reduce to the following familiar transformation groups in the Euclidean plane.

Translations

$$\begin{aligned} x' &= x + h \\ y' &= y + k \end{aligned} \tag{5}$$

Rotations

$$\begin{aligned} x' &= x \cos\theta - y \sin\theta \\ y' &= x \sin\theta + y \cos\theta \end{aligned} \tag{6}$$

Rigid Motions

$$\begin{aligned} x' &= x \cos\theta - y \sin\theta + a \\ y' &= x \sin\theta + y \cos\theta + b \end{aligned} \tag{7}$$

Similarity Transformations

$$\begin{aligned} x' &= r(x \cos\theta - y \sin\theta + a) \\ y' &= r(x \sin\theta + y \cos\theta + b) \end{aligned} \qquad r \neq 0 \tag{8}$$

Further note is made of these transformations. The similarity group (8) reduces, for $r = 1$, to the group (7) of *rigid motions;* and (7) in turn reduces, for $a = b = 0$, to the group (6) of *rotations* and, for $\theta = 0$, to the group (5) of *translations.*

Special note is made of the fact that each of the groups (5) to (8) is a subgroup of the group (4) of linear transformations obtained originally by a *change of coordinates* in the projective plane.[1] Why this is so is quite a mystery. Clearing it up is one of the goals of later theory.

Exercise 10.3

1 Show that each of the sets of transformations (5) to (8) of this last section forms a group. *Hint:* See ([53] pp. 91–93).

2 Show that a distance-preserving transformation which sends two given points $A(4,1)$ and $B(7,4)$ (Fig. 10.8) into the respective points $A'(2,5)$, $B'(-1,8)$ may be obtained by two successive reflections.

3 Show that a rigid motion is either a translation or a rotation. *Hint:* See ([53] pp. 89–91).

10.4 Geometric and algebraic invariants

By a geometric invariant of a group of transformations is meant any property of a figure that is unchanged under all transformations of the group.

For example, an earlier figure (Fig. 10.3) has already shown the effect on triangle $A(2,0)$, $B(4,1)$, $C(2,1)$ of the translation (1) below. Figures 10.9 and 10.10 also show the effect on this same triangle of transformations (2) and (3).

Translation

$$x' = x + 4 \qquad y' = y - 2 \tag{1}$$

[1] See Sec. 10.1.

FIG. 10.8

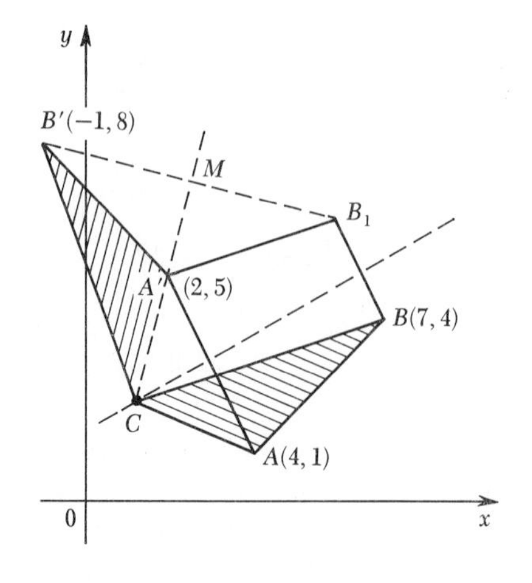

FIG. 10.9

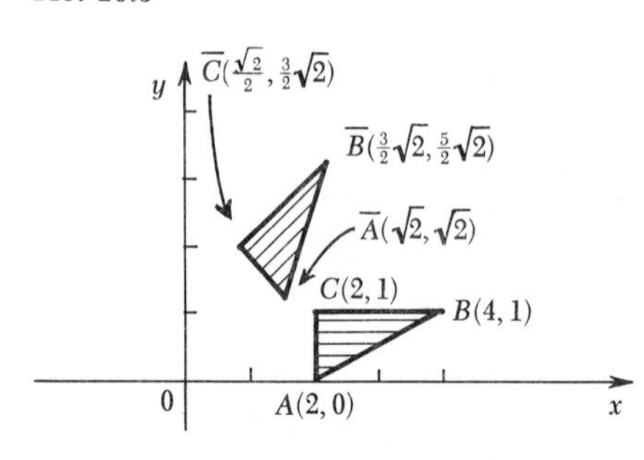

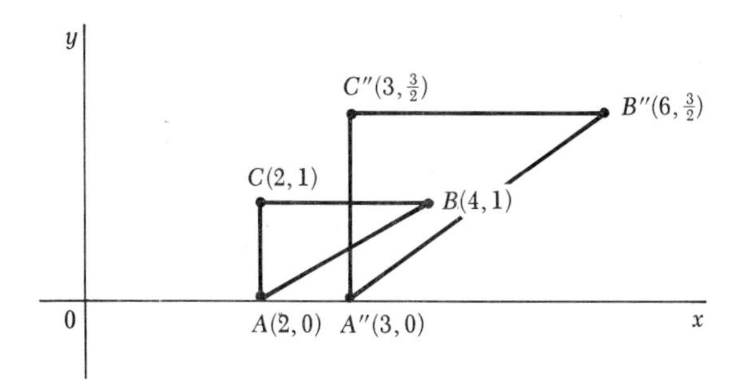

FIG. 10.10

Rotation

$$x' = \frac{x - y}{\sqrt{2}} \qquad y' = \frac{x + y}{\sqrt{2}} \tag{2}$$

Similarity

$$x' = \tfrac{3}{2}x \qquad y' = \tfrac{3}{2}y \tag{3}$$

From Figs. 10.3 and 10.9 it is seen that both size and shape are invariants of (1) and (2), whereas Fig. 10.10 shows that shape, *not size,* is an invariant of (3). Another invariant is defined:

An algebraic expression in the coordinates of points is an invariant of the points under a group of transformations if the expression is unaltered by every transformation of the group.

For example, the expression

$$d^2 = (x_2 - x_1)^2 + (y_2 - y_1)^2$$

representing the distance between the two points $P_1(x_1,y_1)$, $P_2(x_2,y_2)$ of the Euclidean plane is an invariant of the group of rigid motions. The proof is as follows:

From Eq. (7) of Sec. 10.3,

$$\begin{aligned} x_2' - x_1' &= (x_2 - x_1) \cos\theta - (y_2 - y_1) \sin\theta \\ y_2' - y_1' &= (x_2 - x_1) \sin\theta + (y_2 - y_1) \cos\theta \end{aligned}$$

Now, if d' is the distance between the transformed points, squaring, adding, and simplifying these equations gives

$$d'^2 = (x_2' - x_1')^2 + (y_2' - y_1')^2 = (x_2 - x_1)^2 + (y_2 - y_1)^2 = d^2$$

where d' is the distance between the transformed points, and the invariance of d has been shown.

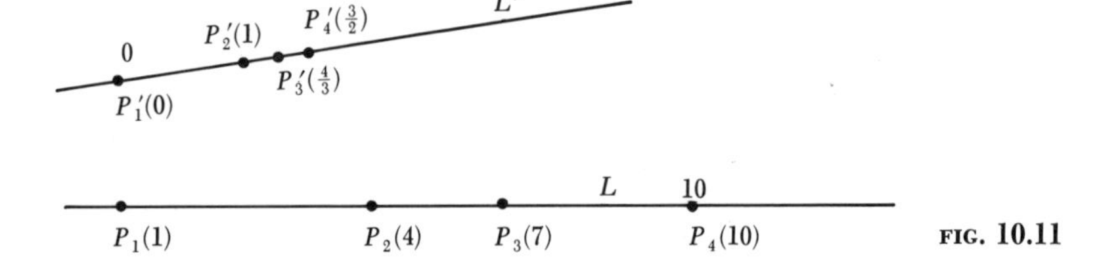

FIG. 10.11

The invariance is verified for distance AB between points $A(2,0)$, $B(4,1)$ and their transformed points $A'(6\text{-}2)B'(8\text{-}1)$ under (1):

$$AB = \sqrt{(4-2)^2 + (1-0)^2} = \sqrt{5}$$

and

$$A'B' = \sqrt{(8-6)^2 + (-1+2)^2} = \sqrt{5}$$

and hence

$$AB = A'B'$$

But distance is *not* an invariant of the next transformation

$$x' = 2\,\frac{x-1}{x+2} \tag{4}$$

which sends the four points

$$P_1(1) \qquad P_2(4) \qquad P_3(7) \qquad P_4(10)$$

of line L (Fig. 10.11) into the four points

$$P_1'(0) \qquad P_2'(1) \qquad P_3'(\tfrac{4}{3}) \qquad P_4'(\tfrac{3}{2})$$

of line L.[1]

Is there an invariant here? The question is answered in what follows.

Exercise 10.4

1 Prove that linear transformations [see (2) of Sec. 10.3] send points on a line into points on a line.

2 a. Show that the translation

$$x' = x + 4 \qquad y' = y - 2$$

[1]See Sec. 10.6.

followed by the rotation

$$x'' = \frac{x' - y'}{\sqrt{2}} \qquad \frac{x' + y'}{\sqrt{2}}$$

gives the transformation

$$x'' = \frac{x - y + 6}{\sqrt{2}} \qquad y'' = \frac{x + y + 2}{\sqrt{2}}$$

b. Name an invariant of this last transformation.

3 Write the transformation which sends point $P(x,y)$ into the point $P'(x,-y)$, and name an invariant of your transformation.

4 Consider the transformation

$$x' - x_0 = (x - x_0) \cos \theta - (y - y_0) \sin \theta$$
$$y' - y_0 = (x - x_0) \sin \theta + (y - y_0) \cos \theta$$

a. Find $(x_0 y_0)$ if these equations transform $A(4,2)$ into $A'(2,4)$, and $B(3,5)$ into $B'(-1,5)$.

b. Name an invariant of your transformation.

5 a. Draw the transform of triangle $A(2,0)$, $B(4,1)$, $C(2,1)$ under the transformation

$$x' = -y + 3 \qquad y' = x + 1$$

b. Is there a property of the figure invariant under the transformation? Explain.

6 a. Is the expression

$$A = x_1 y_2 - x_2 y_1$$

an *algebraic* invariant of transformations (1) and (2) of Sec. 10.2; of (3) of the same section?

b. What does the expression A represent geometrically?

7 The equation

$$x' = \frac{x + 3}{x - 1}$$

represents a projective transformation of a line into itself. Show that it leaves each of the points $x = 3$ and $x = -1$ fixed in position.

8 If a line is carried into itself by the linear transformation

$$x' = \frac{3x + 2}{x + 4}$$

prove that two points A, B of the line remain fixed and that every other point P and its transform divide these fixed points in a constant cross ratio.

9 Find the equation of the projective transformation which carries the points $A(0)$, $B(1)$, $C(2)$ of a line L into the respective points $A'(-1)$, $B'(0)$, $C'(2)$ of a second line L'.

10.5 Linear transformations and projectivities

A very astonishing result is noted now. When line L' of Fig. 10.11 is placed so that $P'(0)$ coincides with $P_1(1)$ as in Fig. 10.12, lines P_2P_2', P_3P_3', P_4P_4' joining corresponding points are concurrent at C. This is not a mere coincidence, as the proofs of Theorems 10.3 and 10.4 will show.

Theorem 10.3 *If L and L' are two lines of the plane, every projectivity from one line to the other is a linear transformation.*

The theorem is proved first when the projectivity is a perspectivity.

Let $O(\mathbf{a})$ be the center of a perspectivity from L to L' (Fig. 10.13), and let $P_0(\mathbf{x}_0)$, $P_1(\mathbf{x}_1)$, and $U(\mathbf{x}_0 + \mathbf{x}_1)$ be the vertices and unit point of a coordinate system on L. Then the line OP_0 meets L' in the point $P_0'(\rho\mathbf{a}_0 + \sigma\mathbf{x}_0)$, or since $\sigma \neq 0$, in the point $P_0'(c_0\mathbf{a} + \mathbf{x}_0)$, with $c_0 = \rho/\sigma$. Similarly, line OP_1 meets L' in the point $P_1'(c_1\mathbf{a} + \mathbf{x}_1)$. Now consider any third point $P(\lambda\mathbf{x}_0 + \mu\mathbf{x}_1)$ on L. The intersection of OP with L' is to be determined. The vector

$$\lambda(c_0\boldsymbol{\alpha} + \mathbf{x}_0) + \mu(c_1\mathbf{a} + \mathbf{x}_1)$$

is a representative vector of a point of L'. But since this vector may also be written

$$(\lambda c_0 + \mu c_1)\mathbf{a} + (\lambda\mathbf{x}_0 + \mu\mathbf{x}_1)$$

it is also a representative vector of a point on line OP. The intersection of OP with L' is therefore the point

$$P'[\lambda(c_0\mathbf{a} + \mathbf{x}_0) + \mu(c_1\mathbf{a} + \mathbf{x}_1)]$$

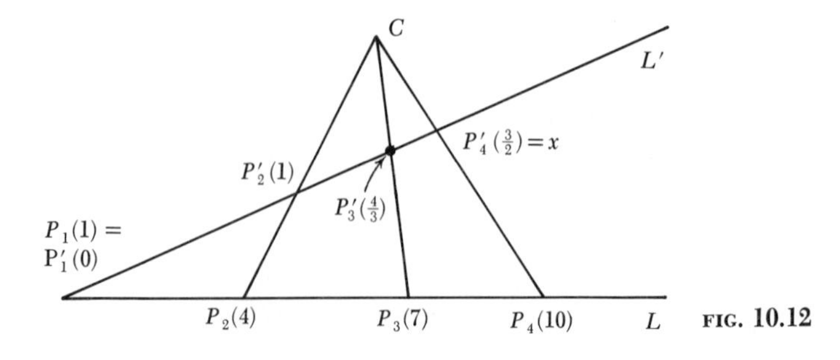

FIG. 10.12

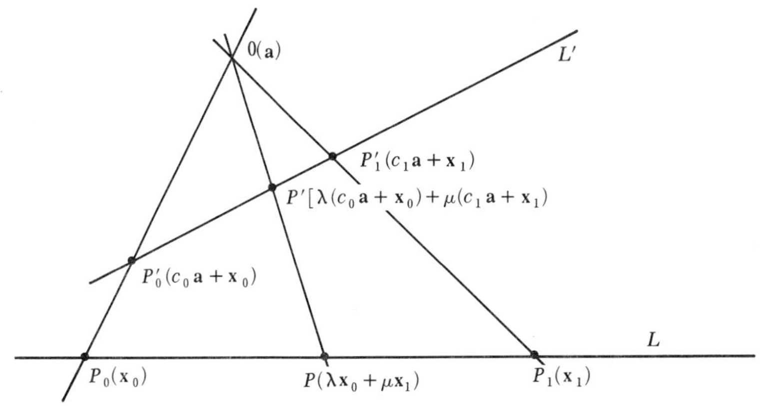

FIG. 10.13

and hence in the coordinate system on L' having P'_0, P'_1 as vertices and $U'(c_0\boldsymbol{\alpha} + \mathbf{x}_0 + c_1\alpha + x_1)$ as unit point, coordinates of P' are (λ,μ). The perspectivity is therefore given by the linear transformation

$$\lambda' = \lambda$$
$$\mu' = \mu$$

or in matrix form

$$\begin{pmatrix}\lambda'\\ \mu'\end{pmatrix} = \begin{pmatrix}1 & 0\\ 0 & 1\end{pmatrix}\begin{pmatrix}\lambda\\ \mu\end{pmatrix}$$

Similarly, by an appropriate choice of coordinate systems on another line L'', a perspectivity from L' to L'' is given by the linear transformation

$$\begin{pmatrix}\lambda''\\ \mu''\end{pmatrix} = \begin{pmatrix}1 & 0\\ 0 & 1\end{pmatrix}\begin{pmatrix}\lambda'\\ \mu'\end{pmatrix} = \begin{pmatrix}1 & 0\\ 0 & 1\end{pmatrix}\begin{pmatrix}1 & 0\\ 0 & 1\end{pmatrix}\begin{pmatrix}\lambda\\ \mu\end{pmatrix} = \begin{pmatrix}1 & 0\\ 0 & 1\end{pmatrix}\begin{pmatrix}\lambda\\ \mu\end{pmatrix} = \begin{pmatrix}\lambda\\ \mu\end{pmatrix}$$

or $\lambda'' = \lambda$ and $\mu'' = \mu$ which is a *linear* transformation.

Since a sequence of perspectivities is a projectivity, the theorem is proved.

To completely identify linear transformations with projectivities, the converse of Theorem 10.3 must also be proved. This is done next.

Theorem 10.4 *Every linear transformation between two lines of the plane is a projectivity.*

proof: Let P_1, P_2, P_3 be three points of line L and P'_1, P'_2, P'_3 be the points of line L' into which they are sent by a linear transformation.

There is a projectivity which takes three given points P_1, P_2, P_3 into three given points P'_1, P'_2, P'_3, and by the theorem just proved, this projectivity is given by a linear transformation. This means that the theorem will be proved if it can be shown that there is *only one linear* transformation establishing the given correspondence.

Coordinate systems are set up on lines L and L' so that P_1, P_2, P_3 are the vertices (1,0), (0,1) and unit point (1,1) on L. The same is done for points P'_1, P'_2, P'_3 on L'. Then the matrices of the transformation taking P_1 and P_2 into P'_1 and P'_2 are of the form

$$\begin{pmatrix} \rho & 0 \\ 0 & \sigma \end{pmatrix}$$

and if, in addition, unit point P goes into unit point P',

$$\rho = \sigma$$

The matrices

$$\begin{pmatrix} \rho & 0 \\ 0 & \sigma \end{pmatrix}$$

all represent the same linear transformation, and thus the theorem is proved.

The arguments just given constitute another proof of the earlier theorem called the fundamental theorem of projective geometry: "A projectivity is completely determined by three points and their corresponding points."

When the arguments just given are extended to 3-space, they yield the theorem:

Theorem 10.5 *Every linear transformation between two planes is a projectivity.*

Dualization of the present theory and extension to 3-space are left as exercises. (See Exercise 10.5, questions 1 and 2 following.)

Exercise 10.5

1 State, illustrate, and prove the dual of Theorem 10.2.

2 Extend Theorem 10.3 to lines of 3-space and dualize. *Hint:* See ([113] chap. 8).

10.6 The invariance of cross ratio on projection

The basic invariant of projective geometry is established in the next theorem. The new proof should be contrasted with the earlier proof given in Sec. 4.13 showing only the invariance of the *absolute* value of the cross ratio.

Theorem 10.6 *Cross ratio is invariant under projection.*

Now that Theorems 10.3 and 10.4 have been proved, Theorem 10.6 will be proved if it can be shown that cross ratio is invariant under the group of linear transformations.

The proof is given for points $P_i(\boldsymbol{\alpha}_i)$, with $i = 1, \ldots, 4$, of line L where components of $\boldsymbol{\alpha}_i$ are x_i, t_i.

Under the linear transformation (1) of Sec. 10.2 the points P_i are transformed into the points $P_i'(A\boldsymbol{\alpha}_i)$, where the matrix A of the system is

$$A = \begin{pmatrix} a_{11} & a_{12} \\ a_{21} & a_{22} \end{pmatrix} \qquad |a_{ij}| \neq 0$$

From (3) of Sec. 9.7 the cross ratio $(P_1P_2P_3P_4)$ is

$$(P_1P_2,P_3P_4) = \frac{\begin{vmatrix} x_3 & t_3 \\ x_1 & t_1 \end{vmatrix} \begin{vmatrix} x_2 & t_2 \\ x_4 & t_4 \end{vmatrix}}{\begin{vmatrix} x_2 & t_2 \\ x_3 & t_3 \end{vmatrix} \begin{vmatrix} x_4 & t_4 \\ x_1 & t_1 \end{vmatrix}}$$

and (P_1',P_2',P_3',P_4') is obtained from the above by replacing each element in the determinants on the right by A times the element. Since such a process does not change the value of the quotient

$$(P_1'P_2',P_3'P_4') = (P_1P_2,P_3P_4)$$

and the theorem is proved.

Verification of the theorem is made for the points P_i and P_i' of Fig. 10.11. Formula (3), Sec. 9.7, yields, under the linear transformation:

$$x' = 2\,\frac{(x-1)}{x+2}$$

$$(P_1P_2,P_3P_4) = \frac{(7-1)(4-10)}{(4-7)(10-1)} = \frac{4}{3}$$

and

$$(P_1'P_2',P_3'P_4') = \frac{(4/3-0)(1-3/2)}{(1-4/3)(3/2-0)} = \frac{4}{3}$$

which shows that

$$(P_1P_2,P_3P_4) = (P_1'P_2',P_3'P_4')$$

Cross ratio is therefore the invariant not revealed by a glance at Fig. 10.11.

By placing line L' in the position shown in Fig. 10.12, one sees graphically how the correspondences established by a linear transformation—and hence by algebraic processes—are established by the more artistic procedure—a perspectivity. Whether one uses algebraic or geometric tools is immaterial, but one must acknowledge the added advantage of the algebraic process which provides coordinates of projected points with a minimum of effort.

Exercise 10.6

1 a. Give homogeneous coordinates of the ideal point I of line L (Fig. 10.11) and of its transform I' on line L' under the transformation (4) of Sec. 10.4.

b. Find (P_1P_2,P_3I) and $(P'_1P'_2,P'_3I')$. Are these cross ratios equal?

2 If the unit of measurement on Euclidean line L (Fig. 10.12) is changed, will lines joining corresponding points still be concurrent at C? Explain.

3 If transformations T_1 and T_2 are given by

$$T_1: \begin{array}{l} kx'_1 = 2x_1 - x_2 \\ kx'_2 = x_1 + x_2 \end{array}$$

$$T_2: \begin{array}{l} lx''_1 = 3x'_1 + x'_2 \\ mx''_2 = x'_1 - x'_2 \end{array}$$

find (a) $T = T_1T_2$ and (b) the transforms of points P_i of Fig. 10.11 under T; and (c) verify the invariance of cross ratio (P_1P_2,P_3P_4) under T.

10.7 Analytic representations of conics

The drawings of Sec. 7.8 showed familiar curves generated by projective pencils of points and others generated by projective pencils of lines. The appearance of these curves, at that time, was a strange inexplicable phenomenon. Now that linear transformations have been identified with projectivities, some of the mystery can be eliminated.

Suppose in the two pencils at S and S' (Fig. 10.14) the reference lines $[a_1,a_2,a_3]$ and $[b_1,b_2,b_3]$ at S correspond to reference lines $[a'_1,a'_2,a'_3]$ and $[b'_1,b'_2,b'_3]$ at S', and suppose unit line at S corresponds to unit line at S'. As lines of the pencil at S and S', they are given by the respective coordinates $[1,0]$, $[0,1]$, and $[1,1]$. Then if line $[y_1,y_2]$ at S corresponds to line $[y'_1,y'_2]$ at S', it follows from the arguments of Sec. 10.5 that the particular projectivity is given by the transformation

$$\begin{array}{l} ky'_1 = y_1 \\ ky'_2 = y_2 \end{array} \tag{1}$$

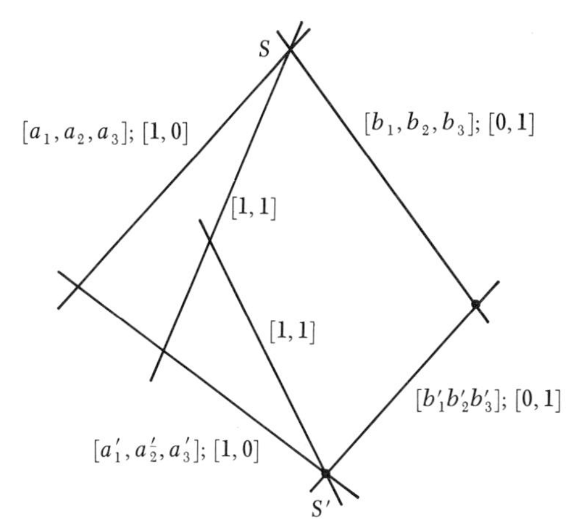

FIG. 10.14

Now coordinates of line $[y_1,y_2]$ are

$$y_1(a_1,a_2,a_3) + y_2(b_1,b_2,b_3)$$

and of line $[y_1',y_2']$ are

$$y_1'(a_1',a_2',a_3') + y_2'(b_1',b_2',b_3')$$

and if point (x_1,x_2,x_3) is to lie on corresponding lines of the pencils, the following equations must be satisfied:

$$\begin{aligned}(y_1a_1 + y_2b_1)x_1 + (y_1a_2 + y_2b_2)x_2 + (y_1a_3 + y_2b_3)x_3 &= 0\\ (y_1'a_1' + y_2'b_1')x + (y_1'a_2' + y_2'b_2')x_2 + (y_1'a_3' + y_2'b_3')x_3 &= 0\end{aligned} \tag{2}$$

or rearranging and using (1),

$$\begin{aligned}y_1(a_1x_1 + a_2x_2 + a_3x_3) + y_2(b_1x_1 + b_2x_2 + b_3x_3) &= 0\\ y_1(a_1'x_1 + a_2'x_2 + a_3'x_3') + y_2(b_1'x_1 + b_2'x_2 + b_3'x_3) &= 0\end{aligned} \tag{3}$$

If solutions y_1, $y_2 \neq 0, 0$ of (3) are to exist, the determinant Δ of (3) must be equal to zero. Hence

$$\Delta = \begin{vmatrix} a_1x_1 + a_2x_2 + a_3x_3 & b_1x_1 + b_2x_2 + b_3x_3 \\ a_1'x_1 + a_2'x_2 + a_3'x_3 & b_1'x_1 + b_2'x_2 + b_3'x_3 \end{vmatrix} = 0$$

Therefore, expanding Δ,

$$\begin{aligned}c_{11}x_1^2 + c_{12}x_1x_2 + c_{13}x_1x_2 + c_{22}x_2^2 + c_{21}x_2x_1 + c_{23}x_2x_3&\\ + c_{33}x_3^2 + c_{31}x_3x_1 + c_{32}x_3x_2 &= 0\end{aligned} \tag{4}$$

where

$$c_{ij} = a_i b'_j - a'_j b_i \qquad i, j = 1, 2, 3$$

By a suitable change of coordinates, (4) may be reduced to the form

$$d_1x_1^2 + d_2x_2^2 + d_3x_3^2 = 0 \tag{5}$$

and in nonhomogeneous coordinates (5) becomes

$$d_1x^2 + d_2y^2 + d_3 = 0 \tag{6}$$

which is recognized as the familiar equation of a conic in the Euclidean plane. Note particularly that the metric concepts used originally in defining conics are missing from the present development.

Of interest later are the special conics

$$x_1^2 + x_2^2 - x_3^2 = 0$$

and the degenerate conic

$$x_3^2 = 0$$

representing the line $x_3 = 0$ counted *twice*.

Dually, the second degree equation in line coordinates u_1, u_2, u_3, that is,

$$au_1^2 + bu_1u_2 + cu_2^2 + du_1u_3 + eu_2u_3 + fu_3^2 = 0$$

with real coefficients, represents a line conic. Special line conics for later use are the conic

$$u_1^2 + u_2^2 + ku_3^2 = 0$$

where k is a constant $\neq 0$, and the degenerate line conic

$$u_1^2 + u_2^2 = 0$$

Exercise 10.7

1 Find the point equation of the conic generated by the projective pencils at $S(0,1,-1)$ and $S'(10,-1,4)$ if reference lines [2,2,2] and [2,1,1] and unit line [4,3,3] at S_1 correspond to reference lines $[-1,2,-3]$ and [1,2,2] and unit line $[0,4,-1]$ at S'. *Hint:* See ([50] pp. 155–157) and question 3 below.

2 Dualize the answer to question 1.

3 Show that the answer to question 1 maybe written in the matrix form

$$(x_1x_2x_3)\begin{pmatrix} 0 & \frac{1}{2} & -\frac{1}{2} \\ \frac{1}{2} & -2 & \frac{3}{2} \\ -\frac{1}{2} & \frac{3}{2} & 1 \end{pmatrix}\begin{pmatrix} x_1 \\ x_2 \\ x_3 \end{pmatrix} = 0$$

10.8 The Erlanger program

As mentioned earlier, one goal of this chapter is an introduction to the Erlanger program, which classifies geometries on the basis of the content or character of theorems rather than on the method used in obtaining them.

The program begins with the formal definition:

> *A geometry is defined by a group G of transformations when its definitions and theorems deal with properties of figures invariant under transformations of G but not invariant under transformations of any other group containing G.*

Thus plane projective geometry is defined by the group of all projective collineations in the plane, since definitions and theorems of this geometry express properties which are invariant under this group. Many of these projective properties are also invariant under more restricted subgroups, such as that of rigid motions. But the system which studies properties invariant under *rigid motion* is Euclidean geometry, and hence, in the Erlanger program, *Euclidean geometry is a special subdivision of projective geometry.* (Strictly speaking, Euclid's elementary system is a mixture of several geometries in the sense of the definition just given. In the theory of similar figures, it takes account only of shape; in the theory of congruent figures, it is concerned also with size. Taken as a whole, it is a study of properties of figures unchanged under rigid motion, for it is only with respect to rigid motion that all properties studied are preserved.)

Concluding remarks

This chapter has been devoted to showing the essential unity of algebra and geometry. It began with the derivation of formulas for change of coordinates in the projective plane. These formulas were then viewed as mappings of points in the plane into points of the same or another plane. Because the formulas were linear in the variables involved, they were called *linear transformations.*

Of chief interest, then, were properties of figures invariant under the *group* of linear transformations, as well as many of its subgroups. Included in the latter were the familiar translations, rotations, and similarity transformations of Euclidean geometry.

Once linear transformations were linked with correspondences established by projectivities, it was possible to establish cross ratio as a fundamental invariant of the linear transformation group and also to obtain equations for curves generated by pencils of points or pencils of lines. The resulting equations explain in part why these curves were the conics of the ancients: ellipses, parabolas, and hyperbolas.

With the aid of transformation groups and their invariants, it was possible to explain how elementary geometric theory was classified under the Erlanger program.

The unification of geometric theory by means of transformation groups and their invariants is one of those monumental projects which has gone far beyond its original objective. When codified under the Erlanger program, the modern science of topology becomes a study of the invariants of topological transformation groups. (By a topological transformation is meant one that is single-valued and continuous and possesses single-valued and continuous inverses.) It should be mentioned, however, that many present-day mathematicians insist that topology is not a branch of geometry, but a self-contained discipline cutting across the fields of both analysis and algebra. Topology brings into algebraic geometry the all-important continuity concept of analysis and clarifies the concept through the employment of geometric language and imagery.

There are other modern studies which for want of a better name are called geometries but which do not permit classification under the Erlanger program, such as the studies of abstract space, initiated by M. Fréchet and others early in the twentieth century. Although these studies do not permit classification under the Erlanger program, they are nevertheless geometric in character. The present material supplies a rich and valuable background for the later study of both topology and abstract space theory.

Suggestions for further reading

Eves, Howard: "A Survey of Geometry," vol. II, chaps. 9 to 11, 16.
Fishbach, W. T.: "Projective and Euclidean Geometry," chaps. 8 and 11.
Graustein, W. E.: "Introduction to Higher Geometry," chaps. 6 and 7.
Hausner, Melvin: "A Vector Space Approach to Geometry," chaps. 8 and 11.
Paige, Lowell J., and J. Dean Swift: "Elements of Linear Algebra," chap. 3.
Rosenbaum, Robert A.: "An Introduction to Projective Geometry and Modern Algebra," chaps. 8 and 9.
Seidenberg, A.: "Lectures in Projective Geometry," chaps. 3, 6 to 8.

11 metric complex projective geometry: other unification theories

Mention has already been made of the fact that a geometry which is based on the real number system is real and a geometry based on the complex number system is complex. After the fashion of algebraists who used complex numbers to extend and generalize their theory, complex geometry will be used now to extend and generalize the geometry of the real plane and to introduce additional unification theories.

Some investigations are made first of the effect of using complex rather than real number triples to represent points or lines of the plane. The totality of points represented by number triples (x_1,x_2,x_3), where x_1, x_2, x_3 are complex numbers, constitutes the complex plane.

A metric is introduced later into this complex plane and the resulting system is metric, complex, projective geometry. Its three main subdivisions are *hyperbolic, elliptic,* and *parabolic* metric geometry. The first two of these are the classical non-Euclidean geometries surveyed in Chap. 3, and the third is the system that so beautifully extends and generalizes Euclid.

Each of these metric systems has its own private, individual measuring stick, and each is providing today's scientists with valuable data for the continuing space explorations of the Jet Age.

11.1 Complex geometry

A brief introduction will be given now to *complex geometry,* in which points and lines are represented by triples of *complex numbers.*

A complex number, it will be recalled, is a number of the form

$$a + bi$$

where a and b are real numbers and $i = \sqrt{-1}$.

If $b = 0$, the complex number is *real,* and if $b \neq 0$, the number is "nonreal". Real numbers are therefore embedded in the complex system, and this fact has been used to extend and generalize algebraic theory. For example, the equation

$$x^2 + 1 = 0$$

which has no solution in the real number field has the two solutions $x = \pm i$, $i = \sqrt{-1}$, in the complex number field.

The simplicity of expression introduced into algebra by extending the real number system has its counterpart in geometry. A simple example will illustrate this statement.

In elementary coordinate geometry, where nonhomogeneous coordinates x, y of a point of the plane are real numbers, the unit circle and the line (Fig. 11.1) are represented by the respective equations

$$x^2 + y^2 = 1 \qquad x = b \tag{1}$$

If $|b| < 1$, the two loci meet at the real points

$$P(b, \sqrt{1 - b^2}) \qquad Q(b, -\sqrt{1 - b^2})$$

If, however, $|b| > 1$, $\sqrt{1 - b^2}$ is a complex number and there is no point of the real plane corresponding to either of the number pairs $(b, \sqrt{1 - b^2})$ and $(b, -\sqrt{1 - b^2})$. Then the circle and the line fail to meet (Fig. 11.2). However, the aim of modern geometers is to set up a geometry in which a conic and a line always meet and have as coordinates a common solution of the equations representing their loci.

There are two ways to accomplish this aim. The real plane may be extended to include points whose homogeneous coordinates are proportional to complex numbers, and existing theory may be modified accordingly; or a geometry of the complex plane may be developed without any reference to the real plane.

The less formal plan of introducing complex geometry as an extension and modification of the real system is contemplated and will suffice here, since complex points and lines will be used only to clarify and extend certain theory of the real plane. (No advanced theory of the complex plane is needed for this purpose.)

FIG. 11.1

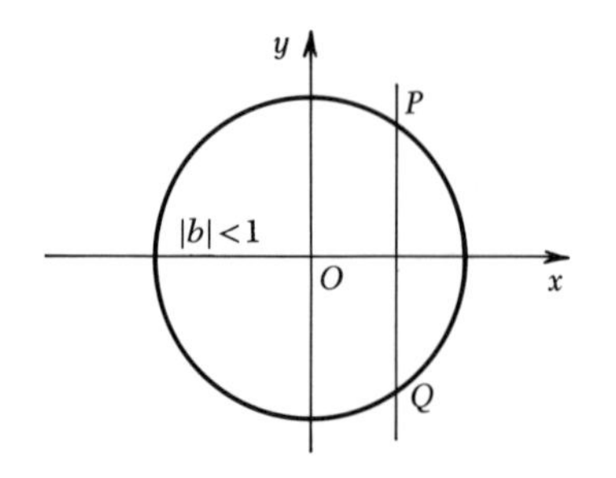

FIG. 11.2

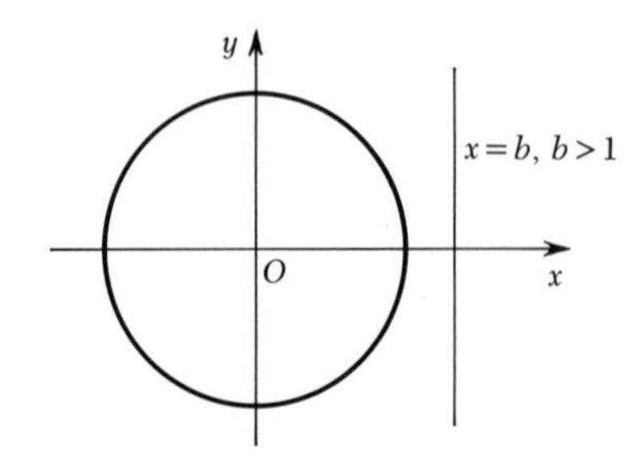

11.2 Complex points

As already shown, the *real* projective plane consists of all equivalent triples with the exception of (0,0,0). To extend this system, consider the set of all equivalent triples of complex homogeneous coordinates (x_1,x_2,x_3). It is agreed that an equivalence triple $\{x_1,x_2,x_3\}$ represents the set of numbers

$$(\rho x_1,\rho x_2,\rho x_3)$$

where ρ is any complex number $\neq 0$. *Each equivalence triple* $\{x_1,x_2,x_3\}$ *is called a complex point, and the totality of these points constitutes the complex plane.*

The complex point $\{x_1,x_2,x_3\}$ is said to be *real* if it has one set of real homogeneous coordinates; otherwise, it is *nonreal.* For example,

$$(2 - 2i,0,4 - 4i) \qquad (2i,0,4i) \qquad (2,0,4)$$

all represent the same real point A. On the other hand, all the number triples

$$(2,0,4i) \qquad (6i,0,-12) \qquad (1,0,2i)$$

represent the same nonreal point B. Both point A and point B are complex.

11.3 Complex lines

Line L of the real plane has already been represented analytically by the linear equation

$$u_1x_1 + u_2x_2 + u_3x_3 = 0 \tag{1}$$

where the number triple (x_1,x_2,x_3) represents a variable point of the line and u_1, u_2, u_3 are constants. Both number triples were assumed to be proportional to a set of real numbers.

This representation of the line is now extended to the complex plane by the assumption that each of the number triples is proportional to a triple of complex numbers. It is then agreed that (1) is the condition that the complex point (x_1,x_2,x_3) lie on the complex line whose line coordinates are u_1, u_2, u_3.

If one set of line coordinates for L is proportional to a set of real numbers, the complex line is *real;* otherwise, it is *nonreal.* For example,

$$2ix_1 + 0x_2 + 4ix_3 = 0 \tag{2}$$

is a real line with line coordinates $\{2,0,4\}$. But

$$2x_1 + 0x_2 + 4ix_3 = 0 \tag{3}$$

is a nonreal line with line coordinates $\{2,0,4i\}$.

If, on the other hand, the point (x_1,x_2,x_3) is fixed and (u_1,u_2,u_3) is a variable number triple, then (1) represents a family of complex lines through the fixed point. It is called the line equation of the point (x_1,x_2,x_3). For example,

$$2iu_1 + 0u_2 + 4iu_3 = 0 \tag{4}$$

is the line equation of the *real* point (2,0,4), $(2i,0,4i)$, and

$$2u_1 + 0u_2 + 4iu_3 = 0$$

is the line equation of the *nonreal* point $(2,0,4i)$.

Other examples of point and line equations are

$$x_3 = 0$$

which is the real line whose line coordinates are [0,0,1] and

$$u_1 + iu_2 = 0$$

which is the nonreal point with coordinates $(1,i,0)$.

11.4 The two-dimensional character of the complex line

Because the algebra of complex numbers is the same as that of real numbers, much of the theory of the real plane carries over immediately to the complex plane. For instance, Eq. (4) of Sec. 8.7 is the point equation of the line through the complex points (a_{21},a_{22},a_{23}) and (a_{31},a_{32},a_{33}), and Eq. (5) of the same section is the line equation of the point of meeting of the complex lines $a[a_{21},a_{22},a_{23}]$ and $b[a_{31},a_{32},a_{33}]$.

Since two complex points determine a line and two complex lines determine a point, incidence axioms for real points and lines hold when these elements are complex. The same is true of the existence axioms (see Exercise 11.1, question 3); but in attempting to extend the theory of the real to the complex plane, a basic difference is noted. In the real plane, the line is a one-dimensional continuum; but in the complex plane, it is a two-dimensional continuum, and separation axioms no longer hold. If the complex nonhomogeneous coordinate λ of point P of a line is written

$$\lambda = x + iy$$

then to each value of λ and hence to each point $P(\lambda)$ of the complex line corresponds the number pair (x,y). The complex line can therefore be brought into 1-1 correspondence with the Gauss (complex) plane, in which each point is represented by the nonhomogeneous coordinates (x,y). Points on the x axis of this Gauss plane correspond to complex points with real coordinates.

Since complex numbers cannot be ordered, separation axioms of the real plane must be denied in the complex plane; in their place is a theory of chains. It is at this point that real and complex geometries take separate paths.

Exercise 11.1

1 Find the equation of the line determined by the points $(1, i, 3 + 2i)$ and $(1, -i, 3 - 2i)$.
2 If the number triples of question 1 denote lines, find coordinates of the point of meeting of these lines.
3 Show that complex points and lines satisfy existence axioms for real points and lines.

11.5 Conjugate complex elements

The complex numbers

$$x = a + bi \qquad \bar{x} = a - bi$$

are *conjugates.* Like numbers, complex points and lines may also be conjugates in accordance with the definition:

> *Two elements are conjugate complex elements if, when (a_1,a_2,a_3) are homogeneous coordinates of one, $(\bar{a}_1,\bar{a}_2,\bar{a}_3)$ are homogeneous coordinates of the other.*

For example, $(1, i, 3 + 2i)$ and $(1, -i, 3 - 2i)$ are conjugate elements, and in this case corresponding coordinates are conjugates. However, because homogeneous coordinates admit a proportionality factor, coordinates of two conjugate complex elements need not be the number triples (a_1,a_2,a_3) and $(\bar{a}_1,\bar{a}_2,\bar{a}_3)$. For example, the number triples

$$(2, i, 1 - i) \qquad \text{and} \qquad (2 + 2i, 1 - i, 2i)$$

represent conjugate complex elements, and yet corresponding numbers are not conjugates. An alternative number triple representing this second element is

$$(2, -i, 1 + i)$$

In the proofs of the following theorems, use is made of the fact that when two conjugate numbers $a + bi$ and $a - bi$ are equal, $b = 0$ and the numbers are real.

Theorem 11.1 *If a point $P(x_1,x_2,x_3)$ lies on a line $L[u_1,u_2,u_3]$, the conjugate complex point $\bar{P}[\bar{x}_1,\bar{x}_2,\bar{x}_3]$ lies on the conjugate complex line $\bar{L}[\bar{u}_1,\bar{u}_2,\bar{u}_3]$.*

proof: The conjugate of the complex number $u_1x_1 + u_2x_2 + u_3x_3$ is the complex number $\bar{u}_1\bar{x}_1 + \bar{u}_2\bar{x}_2 + \bar{u}_3\bar{x}_3$ (see Exercise 11.2, question 1). Also, if a complex number $a + bi$ is zero, $a = b = 0$, and its conjugate $a - bi$ is also zero. Therefore, if

$$u_1x_1 + u_2x_2 + u_3x_3 = 0$$

then

$$\bar{u}_1\bar{x}_1 + u_2\bar{x}_2 + \bar{u}_3\bar{x}_3 = 0$$

But this last equation is the condition that the point $\bar{P}(\bar{x}_1,\bar{x}_2,\bar{x}_3)$ lie on the line $\bar{L}[\bar{u}_1,\bar{u}_2,\bar{u}_3]$, and thus the theorem is proved.

Theorem 11.2 *There is one and only one real point on a nonreal complex line.*

proof: The nonreal complex line L, whose equation is

$$u_1x_1 + u_2x_2 + u_3x_3 = 0$$

may be expressed in the form

$$(u_1' + iu_1'')x_1 + (u_2' + iu_2'')x_2 + (u_3' + iu_3'')x_3 = 0$$

where u_i' and u_i'', with $i = 1, 2, 3$, are all real numbers and at least one of the u_i'' is not equal to zero.

A real number triple (y_1,y_2,y_3) representing a real point will therefore satisfy this equation if

$$u_1'y_1 + u_2'y_2 + u_3'y_3 = 0$$

and

$$u_1''y_1 + u_2''y_2 + u_3''y_3 = 0$$

It is known from the theory of algebra that these equations have the solutions

$$y_1 = k\begin{vmatrix} u_2' & u_3' \\ u_2'' & u_3'' \end{vmatrix} \qquad y_2 = k\begin{vmatrix} u_3' & u_1' \\ u_3'' & u_1'' \end{vmatrix} \qquad y_3 = k\begin{vmatrix} u_1' & u_2' \\ u_1'' & u_2'' \end{vmatrix}$$

where k is a constant $\neq 0$. Hence the nonreal line L contains the real point (y_1,y_2,y_3).

If there were a second real point on line L, the line would be real, contrary to hypothesis. Thus the theorem is proved.

The dual theorem whose proof is left as an exercise is:

Theorem 11.3 *There is one and only one real line on a nonreal point.*

Exercise 11.2

1 Show that the complex numbers $u_1x_1 + u_2x_2 + u_3x_3$ and $\bar{u}_1\bar{x}_1 + \bar{u}_2\bar{x}_2 + \bar{u}_3\bar{x}_3$ are conjugates.

2 Prove the following theorems:
   a. The point of intersection of two conjugate complex lines is a real point.
   b. The line determined by two conjugate complex points is a real line.

3 Find the real point on the nonreal line $[2, i, 3 - 4i]$. [*Hint:* Let (y_1,y_2,y_3) be the real point. Then $2y_1 + 3y_3 = 0$ and $y_2 - 4y_3 = 0$.]

4 Show that the line which joins the nonreal points $(1, i, 3 - 2i)$ and $(1, -i, 3 + 2i)$ is a real line.

5 Find the real point on the nonreal line $(2, i, 3 + 4i)$ and the real line through the nonreal point $(1,i,0)$.

11.6 Real and nonreal conics. The circular points at infinity

Conics too are real or nonreal in accordance with the definition:

> *A conic whose coefficients are real numbers is said to be real if it contains at least one real point; otherwise, it is nonreal.*

For example, the point conic

$$x_1^2 + x_2^2 + kx_3^2 = 0$$

contains, for $k = -1$, the real point $(1,0,-1)$ and hence is real. If $k = 1$, the conic is

$$x_1^2 + x_2^2 + x_3^2 = 0$$

and since there is no real point satisfying this equation, the conic is nonreal. If $k = 0$, the conic is degenerate; and since

$$x_1^2 + x_2^2 = (x_1 + ix_2)(x_1 - ix_2) = 0, \quad i = \sqrt{-1}$$

the conic consists of the conjugate lines

$$x_1 + ix_2 = 0 \qquad x_1 - ix_2 = 0$$

intersecting in the real point (0,0,1).

Dually, the line conic

$$u_1^2 + u_2^2 + ku_3^2 = 0$$

is real if $k = -1$, nonreal if $k = 1$, and degenerate if $k = 0$.

Since

$$u_1^2 + iu_2^2 = (u_1 + iu_2)(u_1 - iu_2) = 0$$

the degenerate line conic consists of the two special conjugate points

$$I(1,i,0) \qquad J(1,-i,0)$$

called the circular points at infinity.

The terminology is due to the fact that if the real conic

$$Ax_1^2 + Bx_1x_2 + Cx_2^2 + Dx_1x_3 + Ex_2x_3 + Fx_3^2 = 0$$

contains these points, then

$$A - Bi - C = 0 \qquad \text{and} \qquad A + Bi - C = 0$$

and hence

$$A = C \qquad B = 0$$

This is the condition that the conic be a circle.

11.7 The Cayley-Klein definitions of distance and angle

A metric is introduced into the complex plane by means of the extremely sophisticated Cayley-Klein definitions for distance and angle in the projective plane. These definitions are stated first and then discussed.

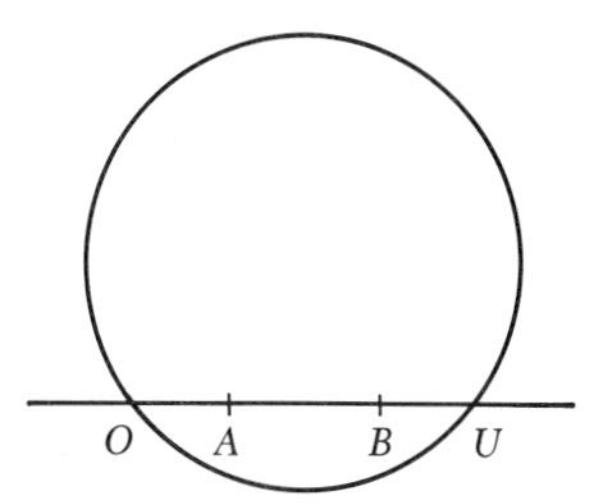

FIG. 11.3

An arbitrary conic of the plane, called the *absolute,* is chosen. Assume for the present that this conic is nondegenerate. Then, any line in the plane not a tangent to the conic meets it in two distinct points, say O and U, and the distance AB between the given points A and B is, by definition, any one of the values

$$AB = c \log (OU,AB)† \tag{1}$$

where c is an arbitrary constant and (OU,AB) is the cross ratio of the four points O, U, A, B (Fig. 11.3).

Dually, any two lines a, b (Fig. 11.4) meet in a point P, and from P there are two tangents o and u to the absolute conic. The angle (ab) between the lines is by definition any one of the values

$$\angle ab = c' \log (ab,ou)$$

where c' is an arbitrary constant and (ou,ab) is the cross ratio of the four lines a, b, c, d.

For the present, no consideration is given the special cases which arise when the points and lines in question are on the conic or are tangent to the conic.

† In the complex plane, the logarithmic function has period $2n\pi$.

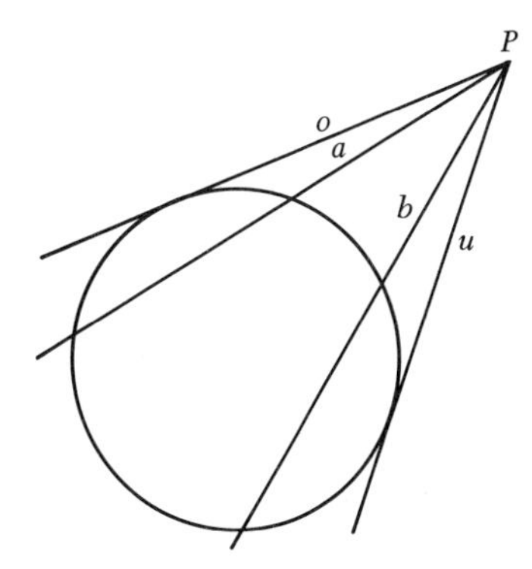

FIG. 11.4

There is not space here for a full and complete motivation of these definitions, but a few preliminary remarks will emphasize some of their significant features. Certainly the many concepts appearing in the definitions, such as an *absolute conic,* a *cross ratio,* a *logarithm,* and an *arbitrary constant,* are confusing, but justifications can be given for all of them. The cross-ratio and logarithm concepts are discussed first.

Just as in elementary geometry distance and angle are invariants of rigid motion, so too in projective geometry it is desired to have distance and angle invariant under projective transformations. It is also desired that formulas for these quantities have Euclidean formulas as special cases.

Although cross ratio has the desired invariance property, it involves four elements, not two, and so two additional points O, U were paired with the points A and B in order to define the projective distance AB. The points O and U are the intersections of line AB with the arbitrarily chosen conic (Fig. 11.3). Thus, distance in projective geometry is not an inherent property of two points but a property of these points in relation to a certain arbitrarily chosen conic.

Furthermore, if A, B, C are three collinear points in the Euclidean plane, the distances AB, BC, AC satisfy the additive relation

$$AB + BC = AC$$

and it is desired to have this property satisfied by the new definition of distance.

But, the cross ratios (OU,AB), (OU,BC), (OU,AC) do not satisfy the additive relation since

$$(OU,AB) + (OU,BC) \neq (OU,AC) \tag{2}$$

They do, however, satisfy the relation

$$(OU,AB) \cdot (OU,BC) = (OU,AC)\dagger$$

and by a well-known theorem of logarithms,

$$\log (OU,AB) + \log (OU,BC) = \log (OU,AC) + 2n\pi$$

Hence, apart from the ambiguity due to the periodicity of the log function, the log of a cross ratio rather than the cross ratio itself has the desired additive property.

The role of the absolute conic and the arbitrary constant which enters into the new definition will be brought out in the illustrations of the next two sections.

†In this connection see Exercise 3 following Sec. 9.7.

11.8 Projective distance illustrated

The projective distance AB between two points A, B of the projective plane will be calculated when the absolute is the special conic given by the equation

$$x_1^2 + x_2^2 - x_3^2 = 0$$

and the two points $A(0,0,1)$, $B(x,0,1)$, $-1 < x < 1$, are on the line

$$x_2 = 0$$

[In the Euclidean plane, this conic is a circle of unit radius having points $O(-1)$ and $U(1)$ as extremities of a diameter (Fig. 11.5).]

Since a set of nonhomogeneous coordinates, or parameters, for the collinear points O, A, B, U are, respectively, -1, 0, x, 1, the cross ratio (OU,AB) is

$$(OU,AB) = \frac{1 - x}{1 + x}$$

and the projective distance AB is then

$$AB = c \log \frac{1 - x}{1 + x}$$

When B is at A, $x = 0$, and hence $AA = 0$. Thus, the new distance from point A to itself is zero, as is desired.

But when B is at U, an unexpected result occurs. By the usual limit processes, as $B \to U$, $x \to 1$, $\log [(1 - x)/(1 + x)] \to -\infty$, and

$$AU = -\infty$$

Similarly, as $B \to O$, $x \to -1$, $\log [(1 - x)/(1 + x)] \to \infty$, and

$$AO = \infty$$

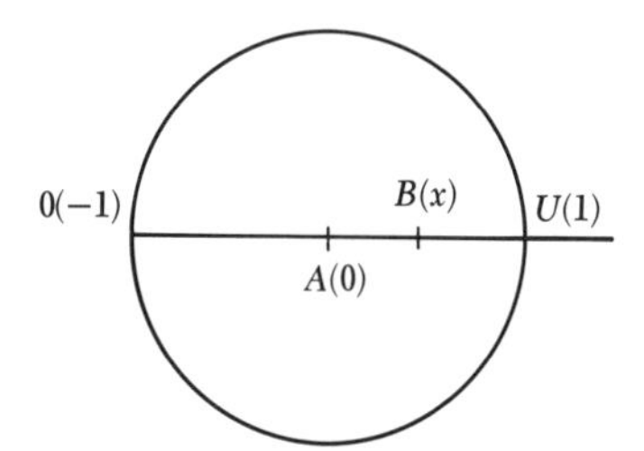

FIG. 11.5

Thus a very surprising result has been obtained: *The projective distance from the center A of a circle to either extremity of the diameter through A is infinite.* (In the Euclidean plane the line segment AO has unit length.)

11.9 Projective angle illustrated

The projective definition of angle is illustrated when the absolute conic is the degenerate conic

$$x_3^2 = 0$$

If two given lines a, b intersect at point P, the tangents from P to this degenerate conic are the lines denoted P_I and P_J, joining P to the circular points at infinity, and the projective angle ϕ between the lines is given by the expression

$$\phi = c' \log (P_I P_J, ab) \tag{1}$$

where c' is an arbitrary constant.

If now P is taken as the origin of a rectangular coordinate system and if the slope m of a line through the two points (x_1,y_1), (x_2,y_2) is a quantity given by the formula

$$m = \frac{y_2 - y_1}{x_2 - x_1}$$

then the slopes of the lines P_I, P_J are, respectively, i and $-i$,

$$(i = \sqrt{-1})$$

Therefore if m_1, m_2 are the respective slopes of lines a, b, the cross ratio $(P_I P_J, ab)$, for convenience denoted r, is given by the expression

$$r = (P_I P_J, ab) = \frac{(m_1 + i)(m_2 - i)}{(m_1 - i)(m_2 + i)} \tag{2}$$

and the Euclidean angle θ between the lines a, b is given by the formula

$$\tan \theta = \frac{m_1 - m_2}{1 + m_1 m_2} \tag{3}$$

or, using (2) and (3),

$$\tan \theta = \frac{1}{i}\left(\frac{r - 1}{r + 1}\right) \tag{4}$$

which, when compared with the well-known relation

$$\tan \theta = \frac{1}{i}\left(\frac{e^{2\theta i} - 1}{e^{2\theta i} + 1}\right) \tag{5}$$

gives $r = e^{2\theta i}$, and hence

$$\theta = \frac{1}{2i} \log (P_I P_J, ab) \tag{6}$$

Thus, for $c' = 1/2i$, the formulas for projective angle and Euclidean angle are alike.

Furthermore, when

$$\begin{aligned} (P_I P_J, ab) &= -1 \\ \log (P_I P_J, ab) &= \pi i \end{aligned}$$

and from (6)

$$\theta = \frac{1}{2i} \pi i = \frac{\pi}{2}$$

The lines a, b are then *perpendicular.*

11.10 A new definition of parallel lines. Hyperbolic, elliptic, and parabolic metric geometries

Since metric projective geometry is primarily concerned with the real points of the plane, a distinction will be made now between the cases when the absolute conic is real and when it is nonreal.

The equation

$$k(x_1^2 + x_2^2) + x_3^2 = 0$$

where k is any real number, represents a real conic if $k < 0$, a degenerate conic if $k = 0$, and an imaginary conic if $k > 0$. Hence, if a line meets the conic in the points O, U, these infinitely distant points are:

Distinct and real if $k < 0$ and the conic is real
Coincident and real if $k = 0$ and the conic is degenerate
Conjugate imaginary if $k > 0$ and the conic is imaginary

A new definition of parallel lines now enters the discussion: *If parallel lines are defined as lines meeting at an infinitely distant (ideal) point of the line,* there will be *two, one, or no* parallels to a line L of the projective plane through a point P not on L according as the absolute conic is *real, degenerate, or imaginary.*

It is at this precise point that projective metric geometry automatically subdivides into three different systems. The geometry which results when there are two infinitely distant (ideal) points on a line, and consequently a right and a left handed parallel to a line, is the classical Bolyai-Lobachevski system called *hyperbolic* geometry; the second metric system, which results when there is only one infinitely distant point on a line, and hence only one parallel to the line, is *parabolic* geometry, the extension of Euclid; and the third metric system, which results when there is no infinitely distant (ideal) point on a line, and hence no parallel to the line, is *elliptic* geometry, the system first developed by Riemann.

(The names hyperbolic, parabolic, and elliptic were proposed by Felix Klein in two monographs appearing in the years 1871 and 1873. The names were suggested to him by the fact that the existence of two real, two imaginary, or two coincident parallels corresponds precisely to the behavior of the asymptotes to the hyperbola, the ellipse, and the parabola.)

11.11 Other classification theories

There is another way in which the complex plane is subdivided. When a real conic, called the absolute conic, has been chosen, the group of all linear transformations leaving the conic invariant is the hyperbolic metric group. The study of properties of figures unchanged under this group is *hyperbolic metric geometry.*

When the absolute conic is nonreal, the group of projective transformations leaving the conic invariant is the elliptic metric group, and the study of properties of figures unchanged under this group is *elliptic metric geometry.*

Occupying an intermediate position between these geometries is the third system, parabolic geometry, in which the absolute conic is degenerate.

The relation between these various systems has been exhibited by means of the character of the points in which a line meets the absolute conic. Since any projective transformation which leaves the absolute unaltered will not disturb the character of metric theorems, the new geometries may also be defined as a *study of properties of figures invariant under the projective transformations that leave the absolute conic invariant.*

To round out the unification theories of the text, the next section will be devoted to showing in what sense parabolic geometry extends and generalizes Euclid.

11.12 Some extensions and generalizations of Euclidean geometry

RIGID MOTIONS

It will be shown first that the familiar transformations of rigid motion which have distance and angle as invariants have a more deeply hidden invariant in the complex parabolic plane. The invariant is given in the next theorem:

Theorem 11.4 *Every rigid motion carries each of the circular points at infinity $I(1,i,0)$, $J(1,-i,0)$ into itself.*

To prove the theorem, it is noted first that, in homogeneous coordinates, Eq. (7) of Sec. 10.3 becomes

$$\begin{aligned} \rho x_1' &= x_1 \cos\theta - x_2 \sin\theta + a x_3 \\ \rho x_2' &= x_1 \sin\theta + x_2 \cos\theta + b x_3 \\ \rho x_3' &= x_3 \end{aligned} \qquad (1)$$

where ρ is a constant $\neq 0$. Their effect on the point $I(1,i,0)$ is seen by substituting the values

$$x_1 = 1 \qquad x_2 = i \qquad x_3 = 0$$

in (1). There result the equations

$$\begin{aligned} \rho x_1' &= \cos\theta - i \sin\theta \\ \rho x_2' &= \sin\theta + i \cos\theta = i(\cos\theta - i\sin\theta) \\ \rho x_3' &= 0 \end{aligned} \qquad (2)$$

which, for $\rho = \cos\theta - i\sin\theta$, are satisfied by

$$x_1' = 1 \qquad x_2' = i \qquad x_3' = 0$$

Hence (1) transforms the point $I(1,i,0)$ into itself. Similarly, the same transformation transforms the point $J(1,-i,0)$ into itself, and the theorem is proved.

Since the circular points at infinity, $I(1,i,0)$ and $J(1,-i,0)$, determine the line $x_3 = 0$, an immediate consequence of this theorem is:

Corollary *The line $x_3 = 0$ is an invariant of rigid motion.*

Now that rigid motions have been seen to have the circular points at infinity, I, J, as invariants, one might reasonably inquire whether this latter property completely characterizes such transformations. In other words, is every projective transformation which has I and J as invariants a rigid motion?

This question cannot be answered in the affirmative. Instead, it will be shown that projective transformations which leave each of the circular points at infinity fixed are transformations of similarity and hence are not necessarily rigid motions.

SIMILARITY TRANSFORMATIONS FROM A PROJECTIVE VIEWPOINT

If the projective transformations (collineations)

$$\begin{aligned} \rho x_1' &= a_{11}x_1 + a_{12}x_2 + a_{13}x_3 \\ \rho x_2' &= a_{21}x_1 + a_{22}x_2 + a_{23}x_3 \qquad |a_{ij}| \neq 0 \\ \rho x_3' &= a_{31}x_1 + a_{32}x_2 + a_{33}x_3 \end{aligned} \tag{3}$$

have $I(1,i,0)$ and $J(1,-i,0)$ as invariants, they carry the line $x_3 = 0$ of these points into the line $x_3' = 0$, and hence

$$a_{31} = a_{32} = 0 \qquad \text{and} \qquad a_{33} \neq 0$$

Division of each of Eqs. (3) by a_{33} then gives

$$\begin{aligned} \sigma x_1' &= a_1x_1 + a_2x_2 + a_3x_3 \\ \sigma x_2' &= b_1x_1 + b_2x_2 + b_3x_3 \\ \sigma x_3' &= \qquad\qquad\qquad\quad x_3 \end{aligned} \tag{4}$$

where, for convenience,

$$\sigma = \frac{\rho}{a_{33}} \qquad \frac{a_{1i}}{a_{33}} = a_i \qquad \frac{a_{2i}}{a_{33}} = b_i \qquad i = 1, 2, 3$$

Since $|a_{ij}| \neq 0$,

$$\begin{vmatrix} a_1 & a_2 \\ b_1 & b_2 \end{vmatrix} \neq 0 \tag{5}$$

and since (4) is to carry each of the points $I(1,i,0)$ and $J(1,-i,0)$ into itself,

$$\begin{aligned} \sigma_1 &= a_1 - ia_2 \qquad & \sigma_2 &= a_1 + ia_2 \\ -i\sigma_1 &= b_1 - ib_2 \qquad & i\sigma_2 &= b_1 + ib_2 \end{aligned}$$

where σ_1 and σ_2 are constants $\neq 0$. Therefore

$$\begin{aligned} ia_1 + a_2 + b_1 - ib_2 &= 0 \\ -ia_1 + a_2 + b_1 + ib_2 &= 0 \end{aligned}$$

and hence

$$a_2 = -b_1 \qquad b_2 = a_1 \tag{6}$$

Substitution of these results in (4) then gives

$$\begin{aligned} \rho x'_1 &= a_1x_1 - b_1x_2 + a_3x_3 \\ \rho x'_2 &= b_1x_1 + a_1x_2 + b_3x_3 \\ \rho x'_3 &= x_3 \end{aligned} \tag{7}$$

which in nonhomogeneous coordinates become

$$\begin{aligned} x' &= a_1x - b_1y + a_3 \\ y' &= b_1x + a_1y + b_3 \end{aligned} \tag{8}$$

But, from (5) and (6),

$$a_1{}^2 + b_1{}^2 \neq 0$$

and hence, if $r^2 = a_1{}^2 + b_1{}^2$, it is possible to set

$$\begin{aligned} \frac{a_1}{r} &= \cos\theta & \frac{b_1}{r} &= \sin\theta \\ \frac{a_3}{r} &= a & \frac{b_3}{r} &= b \end{aligned} \tag{9}$$

and reduce (8) to the form

$$\begin{aligned} x' &= r(x\cos\theta - y\sin\theta + a) \\ y' &= r(x\sin\theta + y\cos\theta + b) \end{aligned} \qquad r \neq 0 \tag{10}$$

But these equations represent similarity transformations in the Euclidean plane. Proof has therefore been given of the following theorem:

Theorem 11.5 *The linear transformations which have the circular points at infinity as invariants are the similarity transformations.*

PARALLELISM AND IDEAL POINTS

When the absolute conic degenerates into the real line $x_3 = 0$ taken twice, its points are all ideal. This line is called, therefore, the *ideal* line of the parabolic plane. All other lines are ordinary. If $x_3 \neq 0$, all points which are not ideal are ordinary.

Since an ordinary line L, of the parabolic plane, meets the degenerate conic (i.e., a line) in a single point, and two ordinary lines which meet at an ideal point are parallel, there is only one parallel to L through an ordinary point P not on L. This is Euclid's parallel axiom.

If now the nonhomogeneous coordinate x of an ordinary point $P(x_1,x_2)$ of the line L is taken to be its distance from a fixed point of the line, the infinitely distant point $P(\infty)$ of the line is its ideal point, and this is the point in which all parallels to L meet L. Thus, in parabolic geometry *two parallel lines always meet,* whereas in Euclidean geometry *two parallel lines never meet.*

It is this failure of parallel lines to meet which is constantly hampering geometric reasoning in the elementary science. Take, for example, the elementary theorem:

> *The interior and exterior bisectors of an angle of a triangle meet the opposite side in points which divide the base internally and externally into the same ratio.*

An exception occurs here when P is the vertex of the isosceles triangle PAB (Fig. 11.6) and the interior bisector of angle P meets the base AB at its midpoint C. Then the exterior bisector of angle P is parallel to the base and hence does not meet it. In the parabolic plane, however, this bisector meets the base at an infinitely distant point $D(\infty)$ called the harmonic conjugate of C with respect to A and B. This means that since the cross ratio

$$(AB,CD) = \frac{AC/CB}{AD/DB} = -1$$

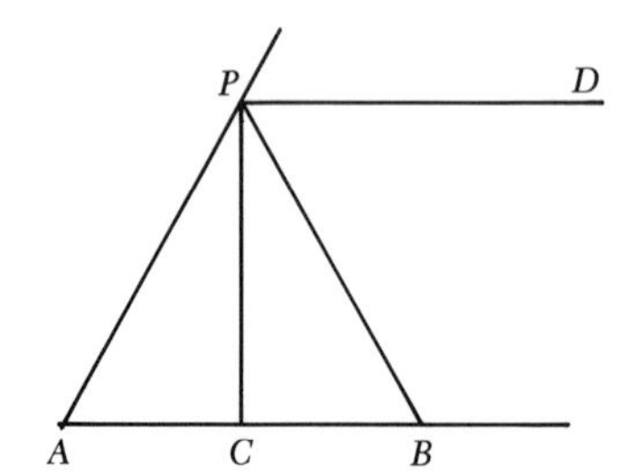

FIG. 11.6

then also $AD/DB = -1$. The ideal point $D(\infty)$ of line AB therefore *divides the line segment AB into the ratio* -1. Use is made of this fact in the following:

A GENERALIZATION OF AN ELEMENTARY THEOREM

Ceva's theorem was used earlier to prove the theorem that the medians of a triangle are concurrent. It is now possible to show that this theorem is only a special case of a much more general theorem of projective geometry. The new theorem is:

Theorem 11.6 *If three distinct points, one on each side of a triangle, are collinear, the three lines which join their harmonic conjugates with respect to the vertices of the triangle to the opposite vertices are concurrent.*

proof: Let A_1, A_2, A_3 (Fig. 11.7) be vertices of the given triangle and let side a_i be opposite the vertex A_i, $i = 1, 2, 3$. Also, let D_1, D_2, D_3 be the three collinear points such that D_i is on side a_i and C_i is the conjugate of point D_i. Then, the theorem will be proved by showing that triangle $C_1C_2C_3$ is perspective to triangle $A_1A_2A_3$. This will necessitate showing that the sides C_3C_2, C_3C_1, C_1C_2 of triangle $C_1C_2C_3$ pass through the respective points D_1, D_2, D_3.

To show, for example, that side C_2C_3 passes through point D_1, project side A_1A_2 from point D_1 on the line A_1A_3. Then points A_1, A_2, and D_3 project into the respective points A_1, A_3, and D_2; and since C_3 is the fourth harmonic point to A_1, A_2, and D_3, C_3 must project into the fourth harmonic point C_2 to A_1, A_3, and D_2. Consequently C_2C_3 passes through D_1. Similarly, side C_1C_3 passes through D_2 and side C_1C_2 through D_3. The triangles $A_1A_2A_3$ and $C_1C_2C_3$ are, therefore, perspective, and hence the lines A_1C_1, A_2C_2, A_3C_3 joining corresponding vertices are concurrent. Thus the theorem is proved.

When line $D_1D_2D_3$ is the ideal line of the plane, point C_i is the midpoint of side a_i, and Theorem 11.6 reduces to the familiar theorem: "The medians of a triangle are concurrent."

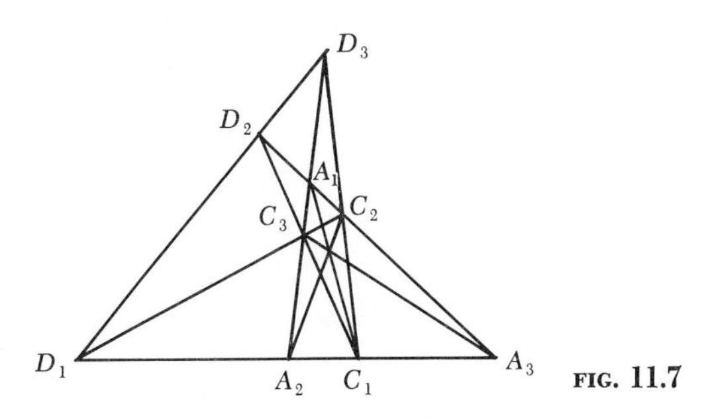

FIG. 11.7

Concluding remarks

This chapter has been devoted to showing what can be accomplished by extending the real projective plane to include complex points and by adding a metric to the resulting set of points. Among other things, the Erlanger classification theories have been extended, and metric complex geometry has been shown to include not only Euclid's metric system but the classical non-Euclidean geometries as well. The view of Euclid from the more lofty projective viewpoint is impressive.

One final remark is in order at this point. A study of current research shows a gradual merging of algebraic geometry with analysis and an increasing use of geometry as a clarifying device in such different fields as the theory of function of a complex variable, linear algebra, statistics, and topology, to mention only a few. The language of geometry is often easier to understand than the language of algebra and analysis: there is economy in both thought processes and the communication of ideas; and often geometric imagery leads to the creation of new mathematical theory.

The trend is marked. *Boundary lines between geometry, algebra, and analysis are growing dimmer with the passage of time.*

Suggestions for further reading

Eves, Howard: "A Survey of Geometry," vol. II, chap. 16, Metric Spaces.
Faulkner, T. E.: "Projective Geometry," chap. 16, Metrical Geometry.
Fishbach, W. T.: "Projective and Euclidean Geometry," chaps. 10 and 11.
Graustein, W. E.: "Introduction to Higher Geometry," chap. 8, Metric Geometry of the Complex Plane.
Robinson, G. de B.: "The Foundations of Geometry," chaps. 5 to 7.
Veblen, O., and J. W. Young: "Projective Geometry," vol. II.

Appendix A

early foundation of Euclidean geometry

1 The definitions of Book I

1 A point is that which has no part.

2 A line is breadthless length.

3 The extremities of a line are points.

4 A straight line is a line which lies evenly with the points on itself.

5 A surface is that which has length and breadth only.

6 The extremities of a surface are lines.

7 A plane surface is a surface which lies evenly with the straight lines on itself.

8 A plane angle is the inclination to one another of two lines in a plane which meet one another and do not lie in a straight line.

9 And when the lines containing the angle are straight, the angle is called rectilineal.

10 When a straight line set up on a straight line makes the adjacent angles equal to one another, each of the equal angles is right, and the straight line standing on the other is called perpendicular to that on which it stands.

11 An obtuse angle is an angle greater than a right angle.

12 An acute angle is an angle less than a right angle.

13 A boundary is that which is an extremity of anything.

14 A figure is that which is contained by any boundary or boundaries.

15 A circle is a plane figure contained by one line such that all the straight lines falling upon it from one point among those lying within the figure are equal to one another.

16 And the point is called the center of the circle.

17 A diameter of the circle is any straight line drawn through the center and terminated in both directions by the circumference of the circle, and such a straight line also bisects the circle.

18 A semicircle is the figure contained by the diameter and the circumference cut off by it. And the center of the semicircle is the same as that of the circle.

19 Rectilineal figures are those which are contained by straight lines, trilateral

[1]Sections 1 to 4 are taken from "The Thirteen Books of Euclid's Elements," 2d ed., a translation from the text of Heiberg with introduction and commentary by Thomas L. Heath. Used by permission of The Macmillan Company for Cambridge University Press, London, 1926.

figures being those contained by three, quadrilateral those contained by four, and multilateral those contained by more than four straight lines.

20 Of trilateral figures, an equilateral triangle is that which has its three sides equal, an isosceles triangle that which has two of its sides alone equal, and a scalene triangle that which has its three sides unequal.

21 Further, of trilateral figures, a right-angled triangle is that which has a right angle, an obtuse-angled triangle that which has an obtuse angle, and an acute-angled triangle that which has its three angles acute.

22 Of quadrilateral figures, a square is that which is both equilateral and right-angled; an oblong that which is right-angled but not equilateral; a rhombus that which is equilateral but not right-angled; and a rhomboid that which has its opposite sides and angles equal to one another but is neither equilateral nor right-angled. And let quadrilaterals other than these be called trapezia.

23 Parallel straight lines are straight lines which, being in the same plane and being produced indefinitely in both directions, do not meet one another in either direction.

2 The postulates

Let the following be postulated:

1 To draw a straight line from any point to any point.

2 To produce a finite straight line continuously in a straight line.

3 To describe a circle with any center and distance.

4 That all right angles are equal to one another.

5 That, if a straight line falling on two straight lines makes the interior angles on the same side less than two right angles, the two straight lines, if produced indefinitely, meet on that side on which are the angles less than the two right angles.

3 The common notions

1 Things which are equal to the same thing are also equal to one another.

2 If equals be added to equals, the wholes are equal.

3 If equals be subtracted from equals, the remainders are equal.

4 Things which coincide with one another are equal to one another.

5 The whole is greater than the part.

4 The forty-eight propositions of Book I

1 On a given finite straight line, to construct an equilateral triangle.

2 To place at a given point (as an extremity) a straight line equal to a given straight line.

3 Given two unequal straight lines, to cut off from the greater a straight line equal to the less.

4 If two triangles have the two sides equal to two sides, respectively, and have the angles contained by the equal straight lines equal, they will also have the base equal to the base, the triangle will be equal to the triangle, and the remaining angles

will be equal to the remaining angles, respectively, namely, those which the equal sides subtend.

5 In isosceles triangles the angles at the base are equal to one another, and, if the equal straight lines be produced further, the angles under the base will be equal to one another.

6 If in a triangle two angles be equal to one another, the sides which subtend the equal angles will also be equal to one another.

7 Given two straight lines constructed on a straight line (from its extremities) and meeting in a point, there cannot be constructed on the same straight line (from its extremities), and on the same side of it, two other straight lines meeting in another point and equal to the former two, respectively, namely, each to that which has the same extremity with it.

8 If two triangles have the two sides equal to two sides, respectively, and have also the base equal to the base, they will also have the angles equal which are contained by the equal straight lines.

9 To bisect a given rectilineal angle.

10 To bisect a given finite straight line.

11 To draw a straight line at right angles to a given straight line from a given point on it.

12 To a given infinite straight line, from a given point which is not on it, to draw a perpendicular straight line.

13 If a straight line set up on a straight line make angles, it will make either two right angles or angles equal to two right angles.

14 If with any straight line, and at a point on it, two straight lines not lying on the same side make the adjacent angles equal to two right angles, the two straight lines will be in a straight line with one another.

15 If two straight lines cut one another, they make the vertical angles equal to one another.

16 In any triangle if one of the sides be produced, the exterior angle is greater than either of the interior and opposite angles.

17 In any triangle two angles taken together in any manner are less than two right angles.

18 In any triangle the greater side subtends the greater angle.

19 In any triangle the greater angle is subtended by the greater side.

20 In any triangle two sides taken together in any manner are greater than the remaining one.

21 If on one of the sides of a triangle, from its extremities, there be constructed two straight lines meeting within the triangle, the straight lines so constructed will be less than the remaining two sides of the triangle, but will contain a greater angle.

22 Out of three straight lines, which are equal to three given straight lines, to construct a triangle: thus it is necessary that two of the straight lines taken together in any manner should be greater than the remaining one.

23 On a given straight line and at a point on it, to construct a rectilineal angle equal to a given rectilineal angle.

24 If two triangles have the two sides equal to two sides, respectively, but have the one of the angles contained by the equal straight lines greater then the other, they will also have the base greater than the base.

25 If two triangles have the two sides equal to two sides, respectively, but have the base greater than the base, they will also have the one of the angles contained by the equal straight lines greater then the other.

26 If two triangles have the two angles equal to two angles, respectively, and one side equal to one side, namely, either the side adjoining the equal angles, or that subtending one of the equal angles, they will also have the remaining sides equal to the remaining sides and the remaining angle to the remaining angle.

27 If a straight line falling on two straight lines make the alternate angles equal to one another, the straight lines will be parallel to one another.

28 If a straight line falling on two straight lines make the exterior angle equal to the interior and opposite angle on the same side, or the interior angles on the same side equal to two right angles, the straight lines will be parallel to one another.

29 A straight line falling on parallel straight lines makes the alternate angles equal to one another, the exterior angle equal to the interior and opposite angle, and the interior angles on the same side equal to two right angles.

30 Straight lines parallel to the same straight line are also parallel to one another.

31 Through a given point to draw a straight line parallel to a given straight line.

32 In any triangle, if one of the sides be produced, the exterior angle is equal to the two interior and opposite angles, and the three interior angles of the triangle are equal to two right angles.

33 The straight lines joining equal and parallel straight lines (at the extremities which are) in the same directions (respectively), are themselves also equal and parallel.

34 In parallelogrammic areas the opposite sides and angles are equal to one another, and the diameter bisects the areas.

35 Parallelograms which are on the same base and in the same parallels are equal to one another.

36 Parallelograms which are on equal bases and in the same parallels are equal to one another.

37 Triangles which are on the same base and in the same parallels are equal to one another.

38 Triangles which are on equal bases and in the same parallels are equal to one another.

39 Equal triangles which are on the same base and on the same side are also in the same parallels.

40 Equal triangles which are on equal bases and on the same side are also in the same parallels.

41 If a parallelogram have the same base with a triangle and be in the same parallels, the parallelogram is double of the triangle.

42 To construct, in a given rectilineal angle, a parallelogram equal to a given triangle.

43 In any parallelogram the complements of the parallelograms about the diameter are equal to one another.

44 To a given straight line, to apply, in a given rectilineal angle, a parallelogram equal to a given triangle.

45 To construct, in a given rectilineal angle, a parallelogram equal to a given rectilineal figure.

46 On a given straight line, to describe a square.

47 In right-angled triangles the square on the side subtending the right angle is equal to the squares on the sides containing the right angle.

48 If in a triangle the square on one of the sides be equal to the squares on the remaining two sides of the triangle, the angle contained by the remaining two sides of the triangle is right.

5 Miscellaneous theorems from other books of Euclid

THE CIRCLE

49 In the same circle or in equal circles, if two chords are equal, they subtend equal arcs.

50 A line through the center of a circle perpendicular to a chord bisects the chord and the arc subtended by it.

51 In the same circle or in equal circles, equal chords are equidistant from the center and chords equidistant from the center are equal.

52 A line perpendicular to a radius at its extremity is tangent to the circle.

53 The tangents to a circle drawn from an external point are equal and make equal angles with the line joining the point to the center.

54 If two circles are tangent to each other, the line of centers passes through the point of contact.

MEASUREMENT OF ANGLES

55 An inscribed angle is measured by half the intercepted arc.

56 An angle formed by two chords intersecting within the circle is measured by half the sum of the intercepted arcs.

57 An angle formed by a tangent and a chord drawn from the point of contact is measured by half the intercepted arc.

58 An angle formed by two secants, a secant and a tangent, or two tangents drawn to a circle from an external point is measured by half the difference of the intercepted arcs.

PROPORTION AND SIMILAR POLYGONS

Definition

Two polygons are similar if corresponding angles are equal, and corresponding sides proportional.

59 The internal (external) bisector of an angle of a triangle divides the opposite side internally (externally) into segments which are proportional to the adjacent sides, and conversely.

60 If two chords intersect within a circle, the product of the segments of one is equal to the product of the segments of the other.

61 If, from a point outside a circle, a secant and a tangent are drawn, the tangent is the mean proportional between the secant and its external segment.

62 If, in a right triangle, the altitude is drawn upon the hypotenuse,

a. The two triangles thus formed are similar to the given triangle and to each other.

b. The altitude is the mean proportional between the segments of the hypotenuse.

c. Each leg of the given triangle is the mean proportional between the hypotenuse and the segment adjacent to the leg.

63 The perimeters of two similar polygons have the same ratio as any two corresponding sides.

AREAS

64 The area of a parallelogram equals the product of its base by its altitude.

65 The area of a triangle equals half the product of its base by its altitude.

66 The area of a trapezoid is equal to half the product of the sum of its bases by its altitude.

67 The areas of two similar polygons are to each other as the squares on any two corresponding sides.

Appendix B
Hilbert's axiom

1 Undefined quantities

A class of undefined elements called *points*, denoted by Latin capitals A, B, C, ...

A class of undefined elements called *lines*, denoted by small Latin letters a, b, c, ...

A class of undefined elements called *planes*, denoted by small Greek letters α, β, γ, ...

Undefined relations: Incidence (being incident, lying on); being in; between; congruence; being parallel; continuous.

2 Axioms of incidence

Incidence is a symmetric relation between elements of different classes (points, lines, planes) such that

2.1 Given any two points A, B, there exists a line a lying on A and B.

2.2 Given A, B, there exists at most one line a lying on A, B.

2.3 There are at least two points which lie on a given line. There are at least three points which do not lie on a line.

2.4 Whenever A, B, C do not lie on a line, there exists a plane α such that A, B, C lie on α. Given any α, there exists a point lying on α.

2.5 If A, B, C do not lie on a line, there is at most one α lying on A, B, C.

2.6 If two points on a line a lie in a plane α, then every point lying on a is also lying on α.

2.7 If two planes have a point in common, then they have at least one more point in common.

2.8 There are at least four points which do not lie on a plane.

3 Axioms of order

3.1 If B is between A and C, then A, B, C are three different points on a line and B is between C and A.

3.2 Let A, C be two points and let AC be the line on which A and C lie. Then, there exists at least one point B on AC such that C is between A and B.

3.3 Let A, B, C be points on a line. Then at most one of them is between the two others.

3.4 (*Pasch's Axiom.*) Let A, B, C be three points which are not on a line. Let b be a line in the plane defined by A, B, C such that none of the points A, B, C lies on b. Let

b intersect the line AB in a point D between A and B. Then, there is either a point X or a point Y on b, where X is between A and C and Y is between B and C.

4 Axioms of congruence

The axioms of incidence and of order make it possible to define the terms "interval," "side of a line or of a plane," "beam or ray," and "angle."

4.1 Let A, B lie on a. Let A' be on a', which may be different from or identical with a. Then it is always possible to find on either side of A' on a' a point B' such that the interval AB is congruent with $A'B'$, i.e., $AB \equiv A'B'$.

4.2 If both $A'B' \equiv AB$ and $A''B'' \equiv AB$, then $A'B' \equiv A''B''$.

4.3 Let AB and BC be two intervals on a which do not have a point in common. Let $A'B'$ and $B'C'$ be two intervals on a' (not necessarily different from a) which also do not have a point in common. Then it follows from

$$AB \equiv A'B' \qquad \text{and} \qquad BC \equiv B'C'$$

that

$$AC \equiv A'C'$$

4.4 Let (h,k) be an angle between the beams h and k which lie on α. Let a' be in a plane α', and let a side of a' in α' be given. Let h' be a beam on α' which starts from O'. Then there exists in α' exactly one beam k' such that $(h,k) \equiv (h',k')$ and that all inner points of (h',k') lie on the given side of a'. Every angle is congruent to itself.

4.5 If two triangles ABC, $A'B'C'$ satisfy $AB \equiv A'B'$, $AC \equiv A'C'$, $\angle BAC \equiv \angle B'A'C'$, then $\angle ABC \equiv \angle A'B'C'$.

4.5* In Axiom 4.5, add the restriction that $\angle ABC = \angle A'B'C'$ provided that AB and $A'B'$ define the right hand beams and that AC and $A'C'$ define the left hand beams of the angles BAC and $B'A'C'$ respectively. Then, we also have $\angle ACB \equiv \angle A'C'B'$.

5 Axiom of parallels

Let a be a line, and let A be a point not on a. Let α be the plane determined by a and A. Then there exists precisely one line on α and A which does not intersect a.

6 Axioms of continuity

6.1 (*Archimedean Axiom.*) If AB and CD are any intervals, then there exist points $A_1, A_2, \ldots, A_n$ on the line AB such that

a. $AA_1 \equiv A_1A_2 \equiv \ldots \equiv A_{n-1}A_n \equiv CD$

b. B is between A and A_n.

6.2 (*Axiom of Completeness.*) The points on a line form a system which cannot be extended if Axioms 2.1, 2.2, 3.1 to 3.4, 4.1 to 4.5, and 6.1 remain valid.

6.2* (*Axioms of Neighborhood.*) If any interval AB is given, then there exists a triangle such that there is no interval congruent to AB in the interior of this triangle.

Bibliography

1 Allendorfer, Carl B., and C. O. Oakley: "Principles of Mathematics," McGraw-Hill Book Company, New York, 1963.

2 Altschiller-Court, N.: "College Geometry," Johnson Publishing Co., Richmond, Va., 1923.

3 Ambrose, Alice, and Morris Lazerowitz: "Logic: The Theory of Formal Inference," Holt, Rinehart and Winston, Inc., New York, 1961.

4 Ayres, Frank: "Theory and Problems of Matrices," Schaum Publishing Co., New York, 1962.

5 Baker, Henry F.: "Principles of Geometry," Cambridge University Press, New York, 1922.

6 Ball, W. W. Rouse: "Mathematical Recreations," 11th ed. revised by H. S. M. Coxeter, The Macmillan Company, New York, 1939.

7 Ballantine, W. B.: "The Logic of Science," Thomas Y. Crowell Company, New York, 1933.

8 Basson, A. H., and D. J. O'Connor: "Introduction to Symbolic Logic," The Free Press of Glencoe, New York, 1960.

9 Bell, E. T.: "The Search for Truth," The Williams and Wilkins Company, Baltimore, 1937.

10 Bentley, A. F.: "Linguistic Analysis of Mathematics," The Principia Press, Inc., Bloomington, Ind., 1932.

11 Birkhoff, G., and S. MacLane: "A Survey of Modern Algebra," The Macmillan Company, New York, 1941.

12 Birkhoff, G.: A Set of Postulates for Plane Geometry Based on Scale and Protractor, *Ann. Math.*, no. 33, pp. 329–345, 1932.

13 Black, Max: "The Nature of Mathematics," Harcourt, Brace and Company, Inc., New York, 1934.

14 Blumenthal, Leonard M.: "A Modern View of Geometry," W. H. Freeman and Company, San Francisco, 1961.

15 Bogoslovsky, Boris B.: "The Technique of Controversy; Principles of Dynamic Logic," Harcourt, Brace and Company, Inc., New York, 1928.

16 Bonola, R.: "Non-Euclidean Geometry," The Open Court Publishing Company, La Salle, Ill., 1912.

17 Borzuk, K., and W. Szmielew: "Foundations of Geometry," Interscience Publishers (Division of John Wiley & Sons, Inc.), New York, 1960.

18 Busemann, Herbert, and Paul J. Kelly: "Projective Geometry and Projective Matrices," Academic Press Inc., New York, 1953.

19 Bussey, W. H.: Finite Geometries, *Trans. Am. Math. Soc.*, vol. 7, pp. 241–259, 1906.

20 Carnap, Rudolf: Foundations of Logic and Mathematics, "International Encyclopedia of Unified Science," vol. I, no. 3, The University of Chicago Press, Chicago, 1939.

21 Carslaw, H. S.: "The Elements of Non-Euclidean Geometry," Longmans, Green & Co., Inc., New York, 1916.

22 Christofferson, H. D.: "Geometry Professionalized," George Banta Company, Inc., Menasha, Wis., 1933.

23 Church, Alonzo: "Introduction to Mathematical Logic," vol. 1, Princeton University Press, Princeton, N.J., 1956.

24 Cogan, E. J.: "Foundations of Analysis," Prentice-Hall, Inc., Englewood Cliffs, N.J., 1962.

25 Cohen, M., and E. Nagel: "Introduction to Logic and Scientific Method," Harcourt, Brace and Company, Inc., New York, 1934.

26 Columbia Associates of Philosophy: "Introduction to Reflective Thinking," Houghton Mifflin Company, Boston, 1923.

27 Cooley, H. R., D. Gans, M. Kline, and H. Wahlert: "Introduction to Mathematics," Houghton Mifflin Company, Boston, 1949.

28 Cooley, H. R., P. H. Graham, F. W. John, and A. Tilley: "College Algebra," McGraw-Hill Book Company, New York, 1942.

29 Committee on the Undergraduate Program, "Elementary Mathematics of Sets," Mathematical Association of America, Buffalo, N.Y., 1958, and Cushing-Malloy Inc., Ann Arbor, Mich., 1958.

30 Copi, I. M.: "Symbolic Logic," The Macmillan Company, New York, 1954.

31 Courant, R., and H. Robbins: "What is Mathematics?" Oxford University Press, Fair Lawn, N.J., 1941.

32 Coxeter, H. S. M.: "The Real Projective Plane," McGraw-Hill Book Company, New York, 1949.

33 Coxeter, H. S. M.: The Affine Plane, *Scripta Math.*, 5–14, 1955.

34 Coxeter, H. S. M.: "Non-Euclidean Geometry," University of Toronto Press, Toronto, Canada, 1947.

35 Coxeter, H. S. M.: "Introduction to Geometry," John Wiley & Sons, Inc., New York, 1961.

36 Daus, Paul: "College Geometry," Prentice-Hall, Inc., Englewood Cliffs, N.J., 1941.

37 Davis, David R.: "Modern College Geometry," Addison-Wesley Publishing Company, Reading, Mass., 1949.

38 Dresden, A.: "An Invitation to Mathematics," Henry Holt and Company, Inc., New York, 1936.

39 Dubnov, Y. A. S.: "Mistakes in Geometric Proofs," translated by A. K. Henn and Olga A. Titelbaum, D. C. Heath and Company, Boston, 1963.

40 Eddington, Sir Arthur S.: "Space Time and Gravitation," Cambridge University Press, New York, 1920.

41 Eddington, Sir Arthur S.: "The Nature of the Physical World," The Macmillan Company, New York, 1928.

42 Enriques, F.: "The Historic Development of Logic," translated by J. Rosenthal, Holt, Rinehart and Winston, Inc., New York, 1929.

43 Exner, R. M., and M. F. Roskopf: "Logic in Elementary Mathematics," McGraw-Hill Book Company, New York, 1959.

44 Eves, Howard: "An Introduction to the History of Mathematics," Holt, Rinehart and Winston, Inc., New York, 1953.

45 Eves, Howard: "A Survey of Geometry," vols. I and II, Allyn and Bacon, Inc., Boston, 1963.

46 Eves, Howard, and C. V. Newsome: "An Introduction to the Foundations and Fundamental Concepts of Mathematics," Holt, Rinehart and Winston, Inc., New York, 1958.

47 Faulkner, T. E.: "Projective Geometry," 2d ed., Interscience Publishers (Division of John Wiley & Sons, Inc.), New York, 1952.

48 Fetison, A. I.: "Proof in Geometry," translated by T. M. Switz and Louise Lange, D. C. Heath and Company, Boston, 1963.

49 Finkbeiner, David T., II: "Introduction to Matrices and Linear Transformations," W. H. Freeman and Company, San Francisco, 1960.

50 Fishbach, W. T.: "Projective and Euclidean Geometry," John Wiley & Sons, Inc., New York, 1962.

51 Forder, H. G.: "The Foundations of Euclidean Geometry," Cambridge University Press, New York, 1927.

52 Fujii, John N.: "An Introduction to the Elements of Mathematics," John Wiley & Sons, Inc., New York, 1961.

53 Graustein, W. C.: "Introduction to Higher Geometry," The Macmillan Company, New York, 1945.

54 Hausner, Melvin: "A Vector Space Approach to Geometry," Prentice-Hall, Inc., Englewood Cliffs, N.J., 1965.

55 Heath, T. L.: "The Thirteen Books of Euclid's Elements," Cambridge University Press, New York, 1926.

56 Heyting, A.: "Intuitionism: An Introduction," North-Holland Publishing Company, Amsterdam, 1956.

57 Hilbert, D.: "The Foundations of Geometry," 6th and 7th eds., The Open Court Publishing Company, La Salle, Ill., 1938.

58 Hilbert, D., and W. Ackerman: "Principles of Mathematical Logic," Chelsea Publishing Company, New York, 1950.

59 Hilbert, D., and S. Cohn-Vossen: "Geometry and the Imagination," Chelsea Publishing Company, New York, 1952.

60 Hodge, W. V. D., and D. Pedoe: "Methods of Algebraic Geometry," Cambridge University Press, New York, 1953.

61 Holgate, T. F.: "Projective Geometry," The Macmillan Company, New York, 1930.

62 Johnson, R. A.: "Modern Geometry," Houghton Mifflin Company, Boston, 1929.

63 Jones, Burton Q.: "Elementary Concepts of Mathematics," The Macmillan Company, New York, 1947.

64 Kemeny, J. G., J. Snell, G. L. Thompson: "Introduction to Finite Mathematics," Prentice-Hall, Inc., Englewood Cliffs, N.J., 1957.
65 Kempe, A. B.: "How to Draw a Straight Line," Macmillan & Co., Ltd., London, 1877.
66 Keyser, C. J.: "Human Worth of Rigorous Thinking," Columbia University Press, New York, 1925.
67 Klein, Felix: "Elementary Geometry from an Advanced Standpoint," The Macmillan Company, New York, 1932.
68 Kline, Morris: "Mathematics in Western Culture," Oxford University Press, Fair Lawn, N.J., 1953.
69 Korzybski, Alfred: "Science and Sanity," 3d ed., The International Non-Aristotelian Library Publishing Co., Lakeville, Conn., 1948.
70 Kuiper, Nicolaas H.: "Linear Algebra and Geometry," North-Holland Publishing Company, Amsterdam; Interscience Publishers (Division of John Wiley & Sons, Inc.), New York, 1962.
71 Langer, Susan K.: "Introduction to Symbolic Logic," Dover Publications, Inc., New York, 1953.
72 Lehmer, D. N.: "An Elementary Course in Synthetic Projective Geometry," Ginn and Company, Boston, 1917.
73 Lewis, C. I., and C. H. Langford: "Symbolic Logic," Appleton-Century-Crofts, Inc., New York, 1932.
74 Lieber, Hugh G., and Lillian R. Lieber: "Non-Euclidean Geometry," Academic Press Inc., New York, 1931.
75 Luneberg, Rudolph K.: "Mathematical Analysis of Binocular Vision," Princeton University Press, Princeton, N.J., 1947.
76 MacLane, Saunders: Metric Postulates for Plane Geometry, *Am. Math. Monthly*, vol. 66, pp. 543–555, 1959.
77 MacLane, Saunders: Symbolic Logic, *Am. Math. Monthly*, vol. 46, pp. 289–296, 1939.
78 MacNeish, H. F.: Four Finite Geometries, *Am. Math. Monthly*, vol. 49, pp. 15–23, 1942.
79 Manning, Henry P.: "Non-Euclidean Geometry," Ginn and Company, Boston, 1901.
80 Meserve, B. E.: "Fundamental Concepts of Geometry," Addison-Wesley Publishing Company, Inc., Reading, Mass., 1955.
81 Minkowski, H.: Time and Space, *Monist*, vol. 28, pp. 288–302, 1918.
82 Moise, E.: "Elementary Geometry from an Advanced Standpoint," Addison-Wesley Publishing Company, Inc., Reading, Mass., 1963.
83 Murdoch, D. C.: "Linear Algebra for Undergraduates," John Wiley & Sons, Inc., New York, 1957.
84 Nagel, E., and J. R. Newman: "Gödel's Proof," New York University Press, New York, 1958.
85 Newman, J. R.: History of Symbolic Logic, in "The World of Mathematics," vol. 3, Simon and Schuster, Inc., New York, 1956.
86 O'Hara, C. W., and D. R. Ward: "An Introduction to Projective Geometry," Oxford University Press, Fair Lawn, N.J., 1937.

87 Nidditch, P. H.: "Elementary Logic of Science and Mathematics," The Free Press of Glencoe, New York, 1960.

88 Paige, Lowell J., and J. Dean Swift: "Elements of Linear Algebra," Ginn and Company, Boston, 1961.

89 Pedoe, Daniel: "A Geometric Introduction to Linear Algebra," John Wiley & Sons, Inc., New York, 1963.

90 Petersen, Julius: "Methods and Theorems for Solutions of Problems of Geometry," Stechert and Co., 1927. Reprinted in "String Figures and Other Monographs," Chelsea Publishing Company, New York, 1960. Original Danish volume published in 1879.

91 Poincaré, H.: "The Foundations of Science," The Science Press, New York, 1913.

92 Poincaré, H.: "Science and Hypothesis," translated by G. B. Halsted, The Science Press, New York, 1905; also Dover Publications, Inc., New York, 1952.

93 Poincaré, H.: "Science and Method," translated by G. B. Halsted, The Science Press, New York, 1913.

94 Poincaré, H.: Review of Hilbert's Foundations of Geometry, *Bull. Am. Math. Soc.*, vol. 10, p. 21, 1903.

95 Pollard, Harry: The Theory of Algebraic Numbers, *Carus Monographs*, no. 9, Mathematical Association of America, Buffalo, N.Y., distributed by John Wiley & Sons, Inc., New York, 1950.

96 Prenowitz, Walter: A Contemporary Approach to Classical Geometry, *Am. Math. Monthly*, vol. 68, Jan, 1961.

97 Prenowitz, W. and M. Jordan: "Basic Concepts of Geometry," Ginn and Company, Boston, 1961.

98 Quine, W. W.: "Methods of Logic," Holt, Rinehart and Winston, Inc., New York, 1950.

99 Ramsey, F. P.: "The Foundations of Mathematics," Kegan Paul, Trench, Trubner & Co., Ltd., London, 1931.

100 Robinson, G. de B.: "The Foundations of Geometry," University of Toronto Press, Toronto, Canada, 1940.

101 Robinson, G. de B.: "Vector Geometry," Allyn and Bacon, Inc., Boston, 1962.

102 Rogers, Hartley, Jr.: An Example in Mathematical Logic, *Am. Math. Monthly*, vol. 70, no. 9, University of Buffalo, Buffalo, N.Y., 1963.

103 Rosenbaum, Robert A.: "Introduction to Projective Geometry and Modern Algebra," Addison-Wesley Publishing Company, Inc., Reading, Mass., 1963.

104 Rosser, J. Barkley: "Logic for Mathematicians," McGraw-Hill Book Company, New York, 1953.

105 Ruskin, John: "The Elements of Drawing and Perspective," John Wiley & Sons, Inc., New York, 1885.

106 Russell, Bertrand: "Introduction to Mathematical Philosophy," 2d ed., The Macmillan Company, New York, 1924.

107 Russell, Bertrand: "An Essay On the Foundations of Geometry," Dover Publications, Inc., New York, 1956.

108 Russell, Bertrand: "Mysticism and Logic and Other Essays," W. W. Norton & Company, Inc., New York, 1929.

109 Russell, Bertrand: "Our Knowledge of the External World," George Allen and Unwin, Ltd., London, 1926.

110 Russell, Bertrand: "Principles of Mathematics," 2d ed., W. W. Norton & Company, Inc., New York, 1937.

111 School, Mathematics Study Group: "Some Basic Mathematical Concepts," vol. I, Yale University Press, New Haven, Conn., 1959.

112 Schreier, D., and E. Sperner: "Introduction to Modern Algebra and Matrix Theory," Chelsea Publishing Company, New York, 1951.

113 Seidenberg, A.: "Lectures in Projective Geometry," D. Van Nostrand Company, Inc., Princeton, N.J., 1962.

114 Shively, L. S.: "Modern Geometry," John Wiley & Sons, Inc., New York, 1939.

115 Smith, D. E., and George Wentworth: "Plane and Solid Geometry," Ginn and Company, Boston, 1913.

116 Smith, P. A., and A. S. Gale: "New Analytic Geometry," Ginn and Company, Boston, 1928.

117 Stabler, E. R.: "An Introduction to Mathematical Thought," Addison-Wesley Publishing Company, Inc., Reading, Mass., 1953.

118 Stabler, E. R.: An Interpretation and Comparison of Three Schools of Thought in the Foundations of Mathematics, *Math. Teacher,* vol. 28, pp. 5–35, 1935.

119 Stoll, Robert R.: "Sets, Logic and Axiomatic Theories," W. H. Freeman and Company, San Francisco, 1961.

120 Struik, Dirk J.: "Lectures on Analytic and Projective Geometry," Addison-Wesley Publishing Company, Inc., Reading, Mass., 1953.

121 Suppes, Patrick: "Introduction to Logic," D. Van Nostrand Company, Inc., Princeton, N.J., 1960.

122 Tarski, A.: "An Introduction to Logic," Oxford University Press, Fair Lawn, N.J., 1939.

123 Taylor, E. H., and G. C. Bartoo: "An Introduction to College Geometry," The Macmillan Company, New York, 1949.

124 Veblen, O., and J. W. Young: "Projective Geometry," vols. I and II, Ginn and Company, Boston, 1910.

125 Veblen, O., and W. H. Bussey: Finite Geometries, *Trans. Am. Math. Soc.,* vol. 7, pp. 241–259, 1906.

126 Watson, Emery Ernest: "Elements of Projective Geometry," D. C. Heath and Company, Boston, 1935.

127 Weyl, Herman: "Philosophy of Mathematics and Natural Sciences," Princeton University Press, Princeton, N.J., 1949.

128 Weyl, Herman: Mathematics and Logic, *Am. Math. Monthly,* vol. 35, pp. 2–13, 1946.

129 Whitehead, A. N.: "An Introduction to Mathematics," Holt, Rinehart and Winston, Inc., New York, 1911.

130 Whitesitt, J. E.: "Boolean Algebra and Its Applications," Addison-Wesley Publishing Company, Inc., Reading, Mass., 1961.

131 Wilder, R. L.: "Introduction to the Foundations of Mathematics," John Wiley & Sons, Inc., New York, 1952.

132 Winger, R. M.: "An Introduction to Projective Geometry," D. C. Heath and Company, Boston, 1923.

133 Wolfe, H. E.: "Introduction to Non-Euclidean Geometry," The Dryden Press, Inc., New York, 1945.

134 Wylie, C. R.: "Foundations of Geometry," McGraw-Hill Book Company, New York, 1964.

135 Yates, R. C.: The Story of the Parallelogram, *Math. Teacher,* vol. 33, 1940.

136 Young, J. W.: "Lectures on Fundamental Concepts of Algebra and Geometry," The Macmillan Company, New York, 1925.

137 Young, J. W.: "Projective Geometry," Carus Monographs, The Open Court Publishing Company, La Salle, Ill., 1930.

138 Young, J. W. A.: "Monographs on Topics of Modern Mathematics," Longmans, Green & Co. Inc., New York, 1911; reprinted by Dover Publications, Inc., New York, 1955.

INDEX